Essentials of Epigenetics

Essentials of Epigenetics

Augustus Drew

FA
FOSTER
ACADEMICS
www.fosteracademics.com

www.fosteracademics.com

FOSTER ACADEMICS

Cataloging-in-publication Data

Essentials of epigenetics / Augustus Drew.
p. cm.
Includes bibliographical references and index.
ISBN 978-1-64646-608-5
1. Epigenetics. 2. Genetics. I. Drew, Augustus.
QH450 .E87 2023
572.865--dc23

Foster Academics,
118-35 Queens Blvd., Suite 400,
Forest Hills, NY 11375, USA

ISBN 978-1-64646-608-5 (Hardback)

Contents

Preface

Epigenetics refers to a field of study concerned with the changes in DNA that do not involve modifications to the underlying sequence. It entails genetic control by factors other than a person's DNA sequence. Epigenetic modifications can activate or deactivate genes and determine which proteins are transcribed. The modifications have various effects on gene expression. The types of epigenetic changes include non-coding RNA, DNA methylation and histone modification. Epigenetic patterns may be affected by a number of lifestyle factors including tobacco smoking, food, exposure to environmental toxins, working on night shifts, obesity, alcohol consumption, psychological stress and physical activity. Epigenetics is a developing area of scientific study that demonstrates how environmental influences such as children's experiences affect the expression of their genes. This book unravels the recent studies in the field of epigenetics. It presents researches and studies performed by experts across the globe. The readers would gain knowledge that would broaden their perspective in this area of study.

This book is a comprehensive compilation of works of different researchers from varied parts of the world. It includes valuable experiences of the researchers with the sole objective of providing the readers (learners) with a proper knowledge of the concerned field. This book will be beneficial in evoking inspiration and enhancing the knowledge of the interested readers.

In the end, I would like to extend my heartiest thanks to the authors who worked with great determination on their chapters. I also appreciate the publisher's support in the course of the book. I would also like to deeply acknowledge my family who stood by me as a source of inspiration during the project.

Editor

1

Introduction to Chromatin

Contents

What You Will Learn in This Chapter

This chapter provides an introduction to chromatin. We will examine the organization of the genome into a nucleosomal structure. DNA is wrapped around a globular complex of 8 core histone proteins, two of each histone H2A, H2B, H3, and H4. This nucleosomal arrangement is the context in which information can be established along the sequence of the DNA for regulating different aspects of the chromosome, including transcription, DNA replication and repair processes, recombination, kinetochore function, and telomere function. Posttranslational modifications of histone proteins and modifications of DNA bases underlie chromatin-based epigenetic regulation. Enzymes that catalyze histone modifications are considered writers. Conceptually, erasers remove these modifications, and readers are proteins binding these modifications and can target specific functions. On a larger scale, the 3-dimensional (3D) organization of chromatin in the nucleus also contributes to gene regulation. Whereas chromosomes are condensed during mitosis and segregated during cell division, they occupy discrete volumes called chromosome territories during interphase. Looping or folding of DNA can bring regulatory elements including enhancers close to gene promoters. Recent techniques facilitate understanding of 3D contacts at high resolution. Lastly, chromatin is dynamic and changes in histone occupancy, histone modifications, and accessibility of DNA contribute to epigenetic regulation.

1.1 Introduction: Epigenetic Regulation in the Context of the Genome

1.1.1 Background: Gene Expression and Chromatin

All organisms inherit traits from their parents, which are encoded in the succession of four bases in nucleic acids. All eukaryotic organisms possess deoxyribonucleic acid (DNA)-based genomes, whereby DNA comprises of antiparallel strands wound in a right-handed double helix. Although the sequence of bases as well as the 3D structure of the DNA helix contribute to the expression of traits, it is thought that the DNA sequence facilitates trait generation by the regulated expression of genes. Gene expression is not the only function of the genome, but replication and faithful inheritance of the genome to descendant somatic cells and the next generation, and evolutionary modifications of the genome are principal functions of the heritable material. Importantly, a genome is not sufficient for generating an organism; this requires a suitable reader that can be represented by a cell or, in the case of higher organisms, an oocyte or zygote that must be from the same species as the genome. The reader implements molecular processes in the cell's nucleus that lead to the production of biomolecules – ribonucleic acids and other biosynthetically active molecules – that replenish the cells and maintain organismal tissues. Through feedback by transcription factors that are encoded in the genome and typically bind to specific DNA sequences, complex regulatory circuits are established. In addition, the DNA itself is organized in a chromatin fiber that facilitates the imposition of information along the DNA sequence. Chromatin also supports the transduction of this epigenetic information into regulatory processes that, in turn, affect transcription, replication, repair, and, in specialized cases, even changes of the DNA sequence. Reciprocal feedback of transcription, combinatorial activity of regulators, genome size, and temporal dynamics contribute to the complexity of gene regulatory networks. On the one hand, this has led to the evolution of developmental programs for complex body plans and, on the other hand, often makes the understanding of individual processes difficult. Therefore, exploration of mechanism of chromatin regulation relies on well suited model systems that disambiguate the function of the components involved. Genomic imprinting and X chromosome inactivation are phenomena where expressed and repressed copies of individual genes are present in the same nucleoplasm and facilitate the study of chromatin-based regulation of transcription factor activity. In addition, ingenious approaches have been designed to analyze chromatin-based heritable expression states controlled by *Polycomb* (PcG) and *Trithorax* group (TrxG) complexes. This chapter contains an introduction to the organization and function of chromatin and provides a basis for understanding its role in regulating the cellular function of DNA sequences.

1.1.2 Discovery of the Nucleosomal Structure of the Genome

Linear arrays of spherical particles of about 70 Ångström in diameter were initially observed by electron microscopy of chromatin released from animal cells (◘ Fig. 1.1) (Olins and Olins 1974). The regular spacing of the spherical units and the fact that similar arrangements were found in many eukaryotes suggested a basic form of organization of the genome. Consistent with a regular structure, experiments using limited digestion of chromatin with nucleases and gel electrophoresis analysis revealed DNA fragments at regular intervals at multiples of 150 base pairs (bp). This is consistent with protection of around 150 bp of DNA from nuclease digestion, whereby nuclease cleavage occurs on the stretch of DNA that lies between spherical particles. Based on these studies, it has become clear that an understanding of the genome needs to consider the molecular components of nucleosomal structure.

1.2 The Structure of the Nucleosome

A nucleosome represents a single repeat unit for organizing the majority of the DNA of a cell.[1] A nucleosome consists of 8 histone proteins and 146 bp of DNA, which is wrapped around them in two left-handed turns[2] (◘ Fig. 1.2a–c). Histone proteins form an octamer complex that comprises positively charged surfaces formed by basic amino acid side chains, which interact with the negatively charged phosphate groups of the DNA backbone. The octamer is assembled from two of each histone H3/H4 and histone H2A/H2B dimers (◘ Fig. 1.2d). The histone H3/H4 dimers occupy the core of the nucleosome and the H2A/H2B dimers are more loosely associated

1 An exception to nucleosomal organization of eukaryotic genomes is found in male gametes of animals, where histones are largely replaced with protamine facilitating a distinct and highly compacted genome configuration.

2 Although it is clear that the large majority of DNA in cells is arranged in left-handed turns around the histone octamer, it has been suggested that right-handed DNA looping can be accommodated by the nucleosomal structure and has experimentally been produced *in vitro*.

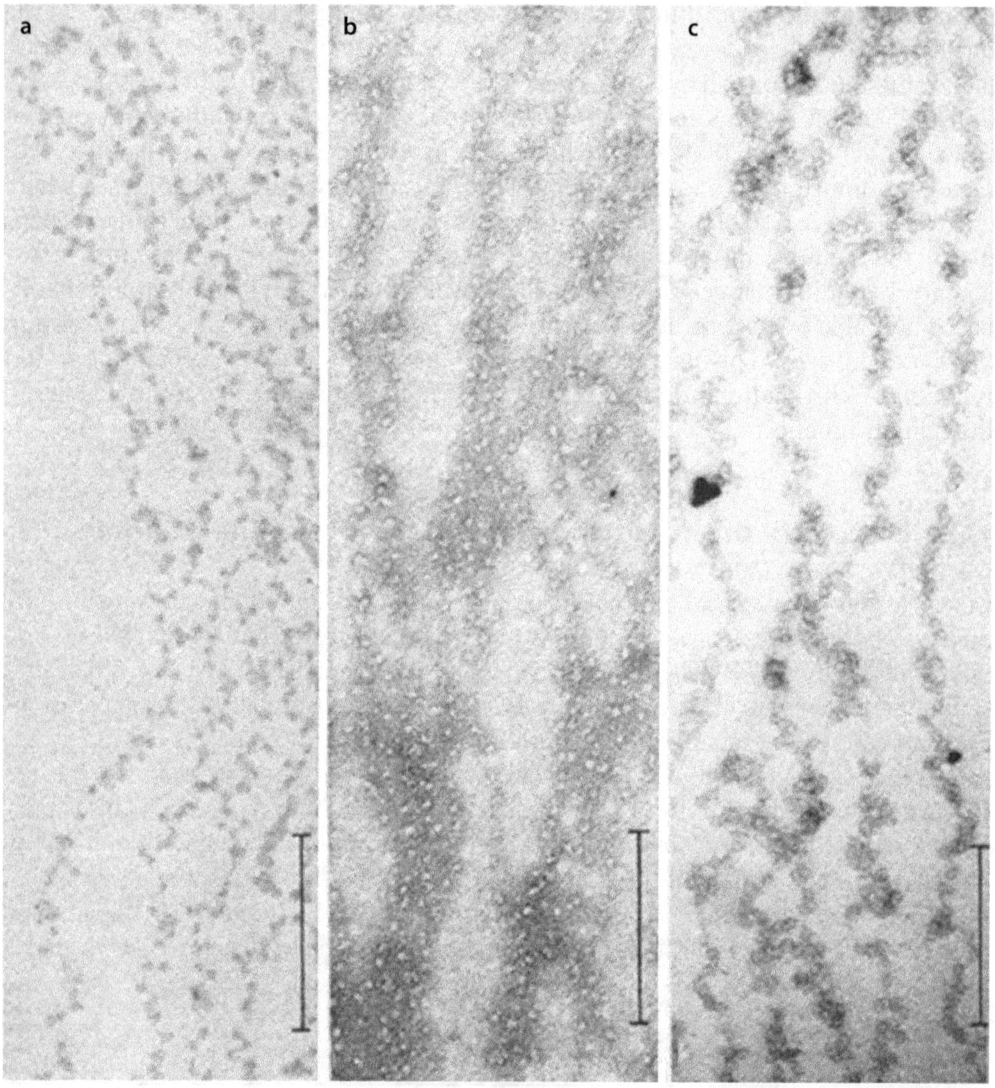

Fig. 1.1 Electronmicrographs of eukaryotic chromatin. Rat thymus chromatin **a** positive and **b** negative staining, and **c** chicken erythrocyte chromatin. (From Olins and Olins (1974))

(Fig. 1.3). Histone proteins form a globular domain with a characteristic alpha-helical arrangement called the histone fold (Luger et al. 1997). Flexible N-terminal regions of histone proteins, the so-called histone tails, associate more loosely with the nucleosome and are accessible for posttranslational modifications.

1.2.1 Histone Variants

Although the nucleosomal structure appears homogenous, variation in histone composition can introduce different functionalities. Different variants of histones can be incorporated, whereby variations of histone H3 and histone H2A are common, and histones H2B and H4 appear to be predominantly canonical.

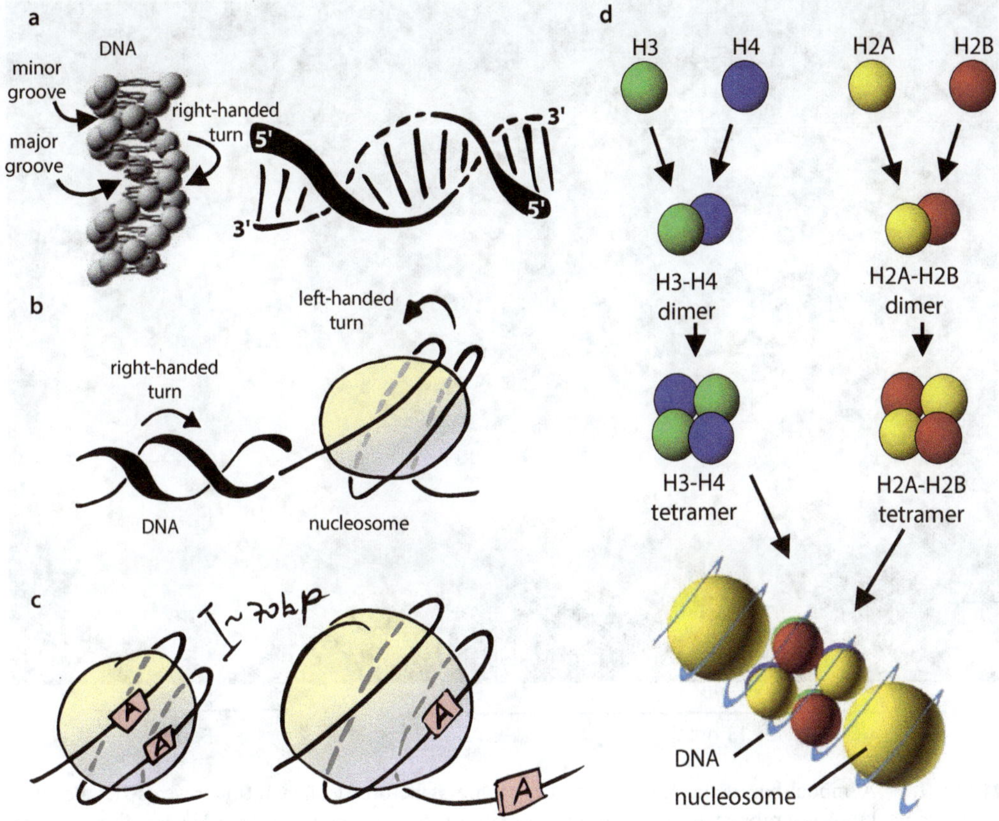

Fig. 1.2 Scheme representing biologically relevant aspects of the nucleosome structure. **a** DNA forms a right-handed double helix with two strands aligned in opposite (antiparallel) orientation. The helix forms a major and minor groove, whereby DNA bases are more accessible from the major groove. **b** Winding of the DNA in left-handed turns over the histone octamere. **c** At a distance of approximately 70 bp, the two DNA loops face the same surface of the nucleosome. Two hypothetical factor binding-sites are indicated by the boxed "A". As one turn of the DNA is completed every 10 bp, both "A" boxes can be facing the surface through the major grove, depending on the position of the nucleosome. Nucleosome positioning along the DNA thereby allows for changes in binding-site geometry that can have a regulatory function. **d** A model for assembly of nucleosomes from histone dimers. DNA assembled into a nucleosomal structure is depicted at the bottom

Histone variants are encoded by a different set of genes and show slight variations in the amino acid sequence. In mammals, histone H3.1 and H3.2 are deposited during DNA replication in S-phase, whereas histone H3.3 is involved in histone exchange at active transcription units and at pericentric and telomeric heterochromatin. Incorporation of histone H3.3 outside of S-phase can be explained as it is the only histone H3 gene whose expression is not restricted to S-phase. Histones are very basic proteins and are associated with chaperones, or loaders, in the cell when they are not in a chromatin context (see book ▶ Chap. 2 of Paro). Different histone loaders are involved in deposition of histone H3 variants. During S-phase, chromatin assembly factor 1 (CAF-1) incorporates histones H3.1 and H3.2 into DNA, whereas *histone cell cycle regulator* (HIRA) has a role in incorporating H3.3 into transcribed

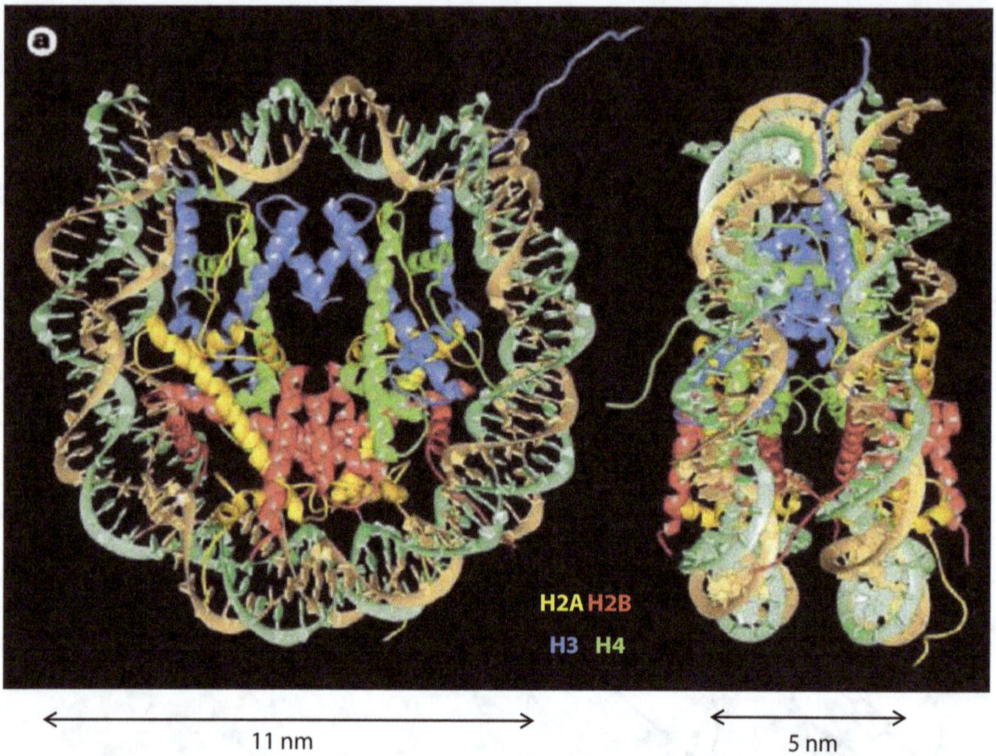

Fig. 1.3 A model-based on X-ray crystallographic structure determination reveals details of a nucleosome. DNA is wrapped around a protein core of 8 histone proteins (histone H3/H4 dimers, green and blue; histone H2AH2B dimers, red and yellow). Globular domains of histone proteins show helical folds and long, unstructured N-terminal peptides are pasted into the model as protruding from the core. (From Luger et al. (1997))

genes. At pericentric[3] and telomeric heterochromatin, ATRX/DAXX is required for histone H3.3 incorporation. At the centromere of mammalian cells, canonical histone H3 is replaced by CENP-A, which contains a number of amino acid differences and adopts a more compact structure. It is thought that CENP-A contributes to the mechanical rigidity of centromeric chromatin that forms the basis of the kinetochore, where spindle microtubules attach for chromosome segregation in mitosis. The specific loader JHURP is involved in CENP-A incorporation into centromeres.

Among histone H2A genes in mammals, the incorporation of the H2A.z variant is restricted to promoters of active genes. This is somehow surprising as H2A.z appears to represent the ancestral form of H2A and is the mammalian H2A gene that is most similar to the H2A gene of *Saccharomyces cerevisiae*. Therefore, the canonical H2A, which is found in the majority of mammalian chromatin, has a slightly different amino acid sequence to the ancestral histone. Additional variants of histone H2A have been correlated with gene activity. Whereas H2A.B is incorporated into chromatin over the transcription unit of active genes, macroH2A accumulates in

3 Pericentic chromatin lies next to the centromere and is normally in a repressed heterochromatic configuration.

silent chromatin. MacroH2A is a vertebrate-specific histone H2A variant that contains a large C-terminal extension, which is commonly referred to as a macro-domain. MacroH2A is enriched at the inactive X chromosome of female mammals, consistent with its association with transcriptionally repressed chromatin (see book ▶ Chap. 4 of Wutz). Through variation of the composition of histone proteins, the function of nucleosomes can be changed, which affects the function of the 146 bp long DNA associated with the nucleosome.

1.3 Histone Modifications

Changes in histone composition are not the only way by which information can be added to nucleosomes. Histone proteins are the subject to posttranslational modifications. In particular, the unstructured N-termini of histones can be modified in a variety of ways, whereby the chemical spectrum of modifications and the combinatorial complexity is high. Phosphorylation of serine, acetylation and methylation of lysine, and methylation of arginine residues are the most prominent histone modifications. Notably, multiple methyl-groups can be added to lysines and arginines, thereby increasing the complexity. Transfer of a ubiquitin to lysines in histone H2A and H2B has also been described. Improvements of analytical techniques have revealed an increasing number of histone modifications and it is likely that this trend will continue in the future. Posttranslational modifications of histones and their functions are incompletely understood at present. However, the development of specific antisera to detect modified histones has provided key insights.

1.3.1 Nomenclature for Histone Modifications

The chemical diversity of modifications and the large number of acceptor sites on histone proteins makes it difficult to describe the modification state of nucleosomes. To facilitate the precise and comprehensive documentation of experimental results and further theoretical elaborations, the scientific community has adopted guidelines for a systematic nomenclature. The class of the histone protein is prefixed to the single character code and position[4] of the amino acid carrying the modification, followed by an abbreviation of the chemical nature of the modification (◘ Fig. 1.4). For example, the short form for histone H3 carrying di-methylation of lysine 27 is H3K27me2. Similarly, H2AK119ub identifies histone H2A carrying a mono-ubiquitin modification on lysine 119. If the precise histone subtype is known this can also be incorporated, for example H3.3K4ac stands for histone H3.3 that is acetylated on lysine at position 4. This versatile nomenclature can easily be extended for multiple modifications: H3K9me3S10p specifies a doubly modified histone H3 with tri-methylation on lysine 9 and phosphorylation on serine 10.

---

4 Numbering of amino acids starts at the biochemical N-terminus, which is different from the methionine encoded by the start ATG of the histone gene due to the activity of methionine aminopeptidase.

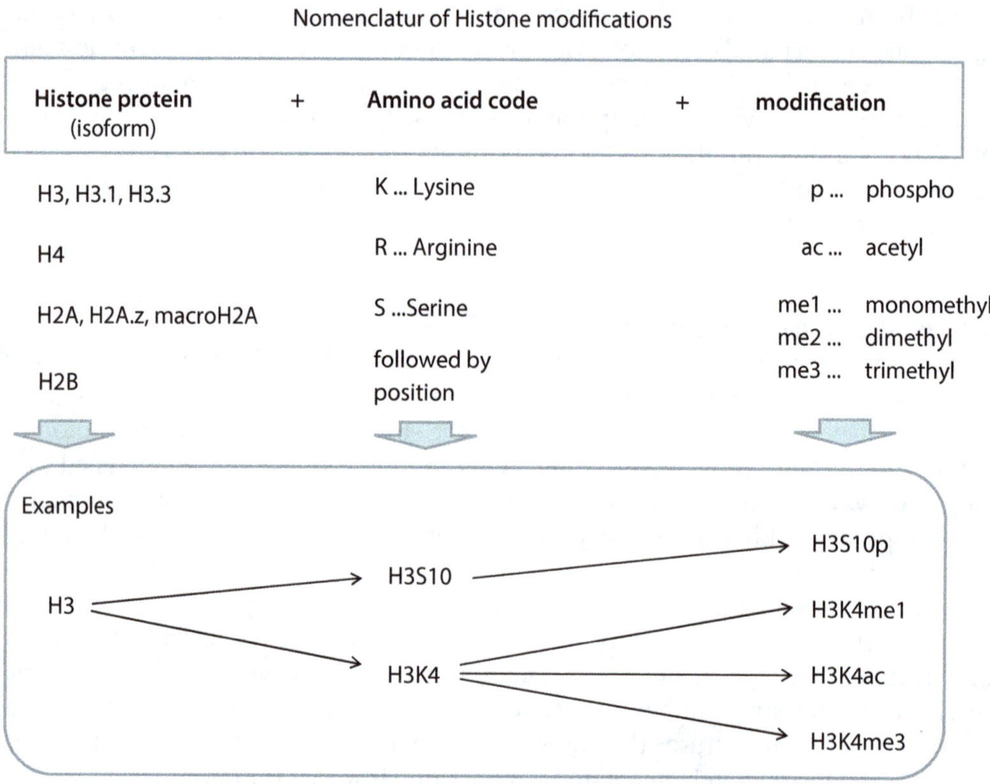

Fig. 1.4 A schematic of short-form nomenclature for posttranslational histone modifications

1.3.2 Combinatorial Modifications at Pericentric Heterochromatin

Histone modifications can be detected by suitable antibodies either using biochemical methods or by staining techniques followed by microscopy. The latter has been used in combination with fluorescence-labeled secondary antibody detection to analyze different histone modification states within the nucleus. Microscopy techniques for dual- or multi-color labeling have further facilitated the observation of co-localization of specific protein complexes with histone modifications. These approaches have been important to characterize different types of chromatin in the nucleus. When combined with fluorescent probes that hybridize with specific DNA sequences, an assessment of the genomic context can be made, albeit at a modest resolution of several thousand bp and, hence, tens or hundreds of nucleosomes.

The pericentric repeats in mouse cells are observed as DNA-dense clusters that are brightly stained with DNA binding dyes such as DAPI.[5] Association of H3K9me3 with this DAPI-dense region suggested a correlation of a specific histone methylation mark with pericentric heterochromatin. Binding of H3K9me3 by the chromo-domain of heterochromatin protein 1 (HP1) leads to HP1 accumulation at the pericentric

5 DAPI (4′,6-diamidino-2-phenylindole) is a fluorescent dye that binds to DNA, allowing the detection of the relative density of DNA in the nucleus by fluorescence microscopy.

heterochromatin. The assay of co-localization with pericentric heterochromatin has been exploited to identify additional components. The curious and instructive anecdote of the human autoimmune serum MCA1 provides a concise understanding. Initial interest in the MCA1 antiserum was raised when it seemed to recognize centromeres in mitosis and, thus, suggested specificity for a potential component of the kinetochore. Since MCA1 did not identify any know kinetochore-associated proteins, mass spectrometry was performed, leading to identification of none but histone proteins. Finally, the mystery was resolved when the specificity of MCA1 was investigated on a panel of doubly modified histone H3 peptides (Hirota et al. 2005). It turned out that MCA1 specifically recognizes H3K9me3S10p, which occurs exclusively at pericentric heterochromatin in mitosis, when serine 10 is phosphorylated by cell cycle kinases, including polo-like kinase (PLK). The physiological effect of serine 10 phosphorylation is to displace HP1 from pericentric heterochromatin in mitosis. This shows that the interaction between HP1 and histone H3 is specific for the H3K9me3 state and is disrupted by doubly modified H3K9me3S10p. This result is important for two reasons. Firstly, it clearly demonstrates that combinations of histone modifications can act in cellular regulation. Secondly, in hindsight, it became clear that antisera for specific histone modifications are sensitive to modifications on neighboring amino acids. This has implications for the interpretation of results obtained with such immunoreagents as a failure to detect a signal does not always correlate with the absence of the modification but can also be caused by interference with a neighboring modification. Although it is assumed that in the majority of cases interference from neighboring modifications can be neglected, there is no systematic study to ascertain the validity of this assumption.

Research on pericentric heterochromatin has also led to the identification of histone methyltransferases. Suppressor of position variegation 3-9 homologues 1 and 2 (Suvar3-9h1 and Suvar3-9h2) associate with pericentric chromatin and catalyze H3K9me3 (◘ Fig. 1.5). H3K9me3 acts as a binding site for HP1 and mediates the recruitment of the additional histone methyltransferases Suvar4-20h1 and Suvar4-20h2 which, in turn, catalyze H4K20me3 at pericentric regions (Schotta et al. 2004). This example shows that modifications from histone H3 can be propagated onto histone H4 within the same chromatin context. The observation that multiple histones and modifications contribute to pericentric chromatin suggests a mechanism for the stability of heterochromatin.

1.3.3 Histone Modifications at High Resolution

Practically, it is of interest to investigate which modifications are present on nucleosomes at a given gene promoter. Knowing of the activity of genes, this can further identify correlations with histone modifications that are generally associated with active or repressed chromatin. For this analysis, two prerequisites need to be fulfilled. Firstly, suitable and specific reagents are required to detect a particular modification of a histone. This has been facilitated by the development of a large array of antibodies that specifically recognize histone modifications. Secondly, the methodology needs to establish a link between modified histones and the DNA sequence, with which these are associated. Chromatin immunoprecipitation (ChIP) does exactly allow for such an analysis (► Method Box 1.1). Chromatin enriched for a defined

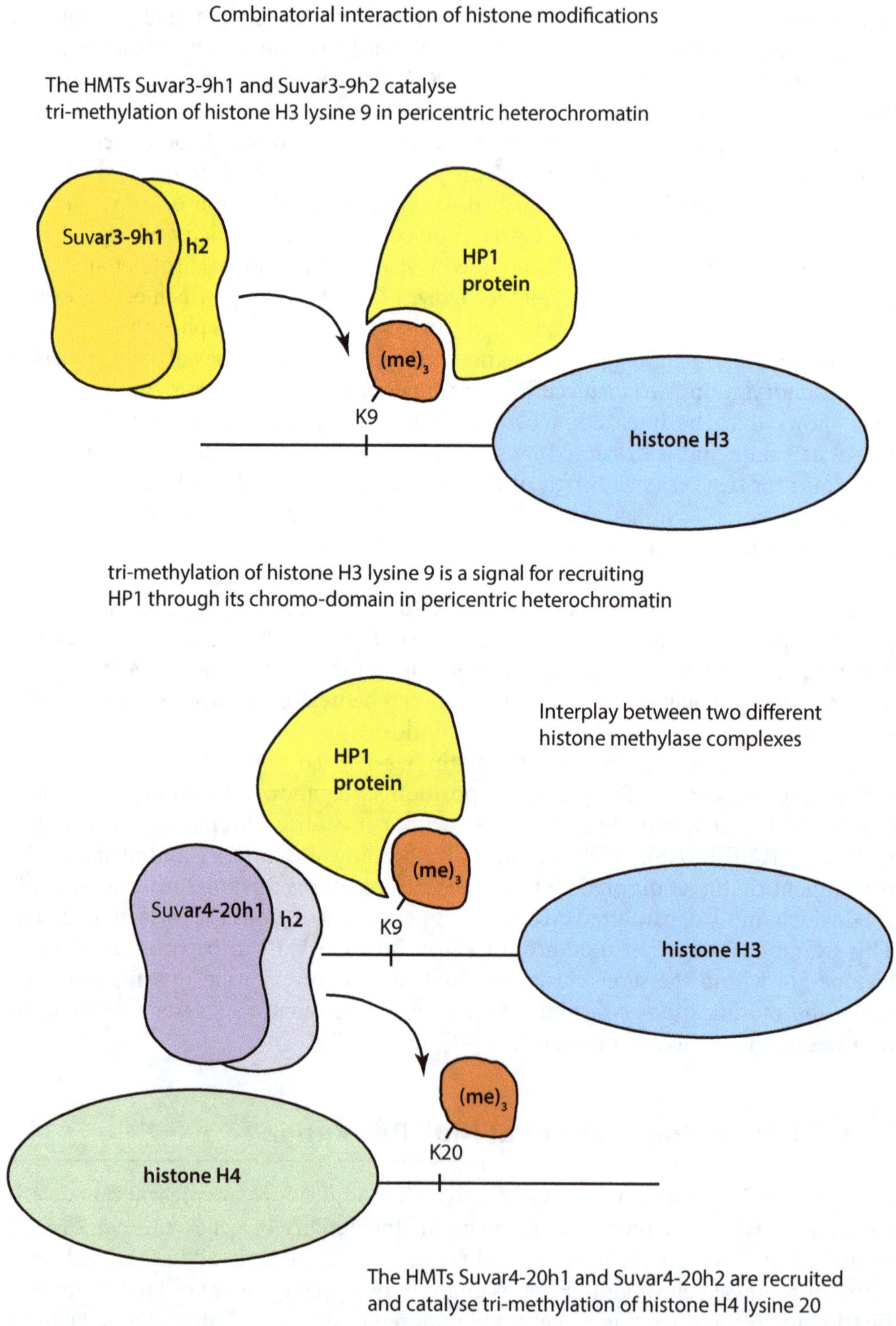

Fig. 1.5 A schematic of histone marks at pericentric heterochromatin. The histone methyltransferases Suv3-9h1 and Suv3-9h2 establish H3K9me3. The tri-methylated lysine 9 serves as a binding signal for HP1 to histone H3. HP1 binding and recruitment of the histone-methyltransferases Suv4-20h1 and Suv4-20h2 leads to establishment of H4K20me1 (Schotta et al. 2004)

histone modification is isolated through immunoprecipitation using antibodies that recognize specific histone modifications. The DNA can then be purified and analyzed by a variety of methods including hybridization to microarrays (ChIP-array), adapter ligation and sequencing (ChIPseq), PCR with gene-specific primers, or by hybridization on dot blots. The latter has originally been used to investigate highly repeated parts of the genome, such as the pericentric repeats, whilst the former can provide genome-wide maps of histone modifications at nucleosomal resolution.

Method Box 1.1: Chromatin Immunoprecipitation

Histone modifications, and transcription factor binding, can be analysed at high resolution by chromatin immunoprecipitation (ChIP). For ChIP analysis, the DNA is cross-linked to histone proteins using formaldehyde, when chromatin is in its native form within cells. Subsequently, cells are lysed and chromatin is sheared either by ultrasound treatment or partial digestion with nucleases. Chromatin that is associated with a specific histone modification or protein is enriched by immunoprecipitation with a specific antibody. After purification, the chromatin associated DNA can be analysed by dot blot, gene specific PCR, and next generation sequencing (◘ Box Fig. 1.1).

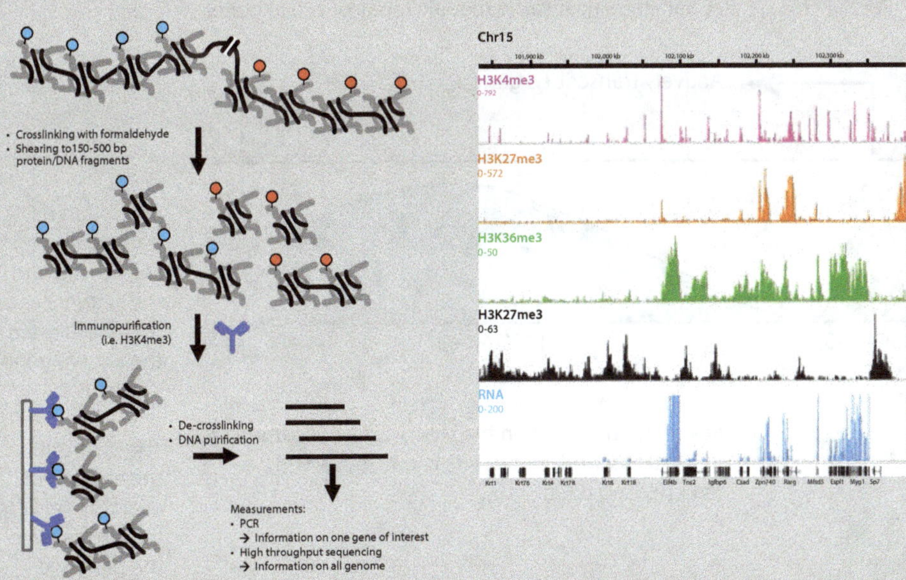

◘ **Box Fig. 1.1** Analysis of chromatin modifications by ChIP. A schematic shows an overview of the ChIP analysis (left) and an example view from a genome browser of a histone modification map obtained by next generation sequencing (right)

1.3.4 Chromatin Modifications Associated with Transcription Units

From analyses in multiple cell types, correlation of gene activity and histone modifications along transcription units has been performed. Some general rules can be deduced. Gene promoters are frequently marked by H3K4me3 and carry acetylated lysines on histone H4 and H3. The gene body of active genes is often marked by H3K36me3. The occurrence of H3K4me3 at the promoter and H3K36me3 over the gene body is a reliable indicator for transcription and has been used to identify new genes (◘ Fig. 1.6) whose transcripts would have been difficult to detect due to their low abundance (Guttman et al. 2009).

In contrast, genes that are characterized by H3K9me2 or H3K9me3 tend to be transcriptionally inactive. Similarly, H3K27me3 and H2AK119ub are associated with the activity of PcG complexes that are well-known repressors of developmentally regulated genes (see book ► Chap. 3 of Paro). In progenitor cells and embryonic stem cells (see book ► Chap. 7 of Paro), these marks can also co-occur with the active mark H3K4me3. This observation has led to the discovery of bivalent chromatin at promoters with apparently active H3K4me3 and repressive H3K27me3 modifications. It appears that this configuration is resolved when cells differentiate and yields active H4K4me3 or repressed H3K27me3 configurations in separate cell

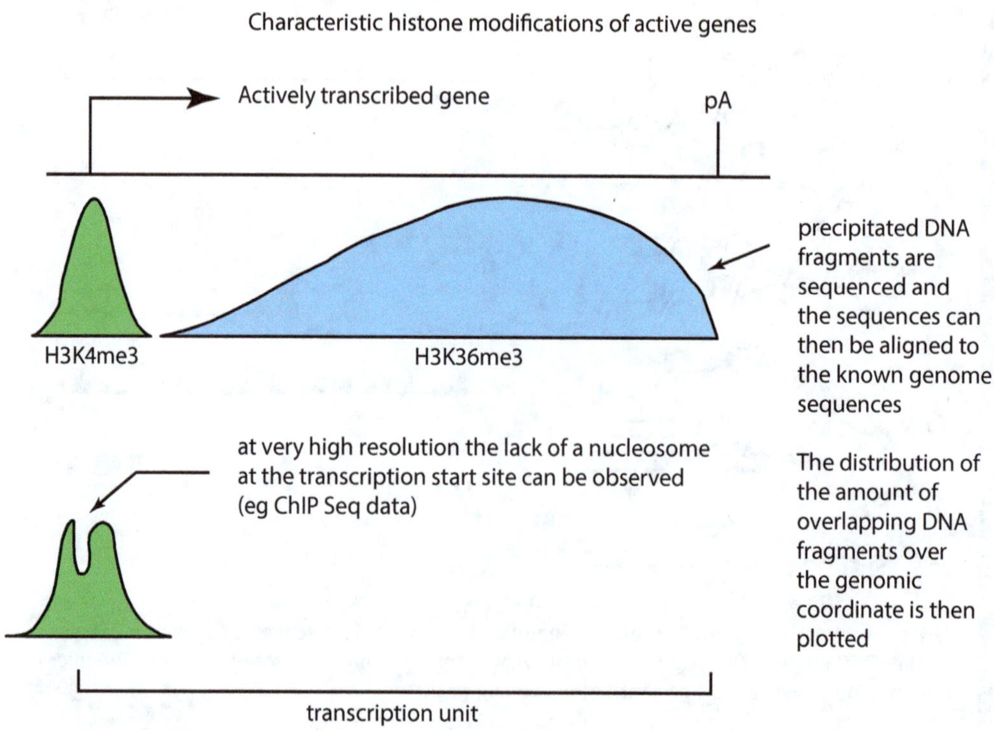

◘ **Fig. 1.6** Illustration of ChIPseq analysis of the transcription unit of a hypothetical active gene. A peak of H4K4me3 enrichment overlaps the gene promoter whereas the transcription unit shows increasing H3K36me3 towards the 3′ end of the gene. If data is obtained at very high resolution, a gap in the H4K4me3 peak can be observed at the position of the transcription start site (TSS). At TSS one nucleosome is displaced and the DNA is bound by general transcription factors

lineages. Therefore, PcG complexes might pre-mark certain developmentally regulated genes before an expression state is committed, which is consistent with the developmental plasticity of progenitor cells (see book ▶ Chaps. 3 and 7 of Paro).

1.3.5 A Concept of Writers, Readers, and Erasers of Histone Modifications

Chemical reactions that lead to posttranslational modifications of proteins use cofactors from metabolic pathways. Acetyl-Co-enzyme A, S-adenosyl-methionine, and ATP are used for the acetylation, methylation, and phosphorylation of histones, respectively (see book ▶ Chap. 9 of Santoro). The corresponding enzymatic activities are referred to as histone acetyltransferases (HATs), histone methylases (HMTs), and histone kinases. Analogously, histone ubiquitin-transferases catalyze monoubiquitinylation. Posttranslational modifications of proteins are frequent and can have different functions. Not all modifications do possess an apparent physiological function and might be observed, to varying extents, as likely bystander reactions that the cell could not prevent. In a number of cases, highly specific modifications have been selected during evolution as signals that can have profound effects on the function of chromatin. Several histone modifications have been implicated in the transcription of genes. Other modifications act to establish a heritable signal for repression. Highly specific functions have been identified for particular methylation states of lysines in the N-termini of histone H3, whereas acetylation can be less specific. Acetylation of lysines of histones is part of the chromatin assembly process and is thought to remove the positive charge from lysines, thereby preventing strong interactions with negatively charged phosphate groups of the DNA. Acetylation of lysines of histone H3 and H4 is also associated with active promoters and could similarly loosen the association of histones with DNA to facilitate accessibility. These observations suggest a charge effect of acetylation and biophysical mechanisms involved in regulating local chromatin accessibility. However, this is not the entire story.

Bromodomain[6]-containing proteins can bind to acetylated lysines and attract other regulators. Proteins that specifically bind modified histones are considered readers of histone modifications. This group of proteins is particularly important for recognizing specific methylations states of lysines in the N-terminus of histone H3. HP1 protein can associate with H3K9me3 through a chromodomain. A single interaction provides only little binding energy given the small size of the tri-methyl-modification. However, HP1 can bind through a cooperative binding mechanism that enhances the binding of several HP1 proteins to longer stretches of H3K9me3 modified chromatin. This explains the enrichment of HP1 on the pericentric regions of mouse chromosomes, where H3K9me3 is abundant. Similar to HP1, the Polycomb (Pc) protein contains a chromodomain which is specific for H3K27me3 (see book ▶ Chap. 3 of Paro). The relevance of chromodomains in cells has been confirmed by replacing the chromodomain of Pc by the one from HP1, whereby the protein was

6 Bromo- and chromodomains are defined based on sequence homology in a number of chromatin-associated proteins. They recognize posttranslational modifications and sequence variation of histones in the context of the nucleosome.

redirected to pericentric chromatin. Similar protein domains exist for reading phosphorylated serines on histones. Importantly, 14-3-3 proteins have high specificity for the doubly modified histone H3S10pK14ac. This specificity plays a role in gene activation. A switch from an inactive H3K9me3 and HP1-bound state to an active H3K9me3S10pK14ac state has been described during the activation of the cell cycle inhibitor p21, which is an important tumor suppressor gene (see ▶ Chap. 8 of Santoro). Histone H3 serine 10 phosphorylation by ERK kinase activity interferes with HP1 binding to H3K9me3 and simultaneously facilitates the association of 14-3-3 proteins if lysine 14 is acetylated. This mechanism demonstrates how cell signaling, in combination with combinatorial histone modifications, can contribute to complex gene regulatory mechanisms.

A last aspect of histone modifications is their stability. Depending on their chemical nature, histone modifications possess different lifetimes. Whereas phosphorylation is readily reversible through phosphatases, tri-methylated lysine modifications can persist for extended periods. Histone deacetylases (HDACs) and histone demethylases (HDMs, or KDMs for lysine demethylases) remove acetyl- and methyl-groups from lysines, respectively. In addition, deubiquitinating enzymes contribute to the turnover of ubiquitin moieties. To date, mechanisms for the enzymatic removal of all histone modifications have been described, suggesting that posttranslational modifications of chromatin are dynamic and actively regulated by the cell. Proteins or complexes that remove histone modifications are considered erasers of epigenetic information. Whereas some histone modifications have been selected by evolution for a regulatory function other may be less important. Examples for the function of histone modifications will be discussed in their physiological context in the following chapters of this book.

1.4 DNA Modifications

Modifications of histones illustrate how information can be added to the genome without changing the sequence of the DNA. This is exactly how we defined "epigenetics" in the opening of the book. Enzymatic activities for establishing and removing modifications also illustrate how this information can be dynamically regulated during development or in response to external stimuli. However, we also had one expectation that is less easily explained: How can the potentially complex patterns of histone modifications be transmitted through cell division? Upon replication of the DNA, twice the number of nucleosomes will need to be assembled and this necessitates that half of the histones are freshly produced whereas the other half keeps the previously established modification patterns (see book ▶ Chap. 2 of Paro). How can the information be reestablished on the new histones? Or is it not restored but lost? There are good indications that epigenetic modifications are maintained but the mechanisms appear to be complex and are poorly understood. To understand the problem of epigenetic heritability, we turn to a much simpler system where maintenance is mediated by a concise mechanism.

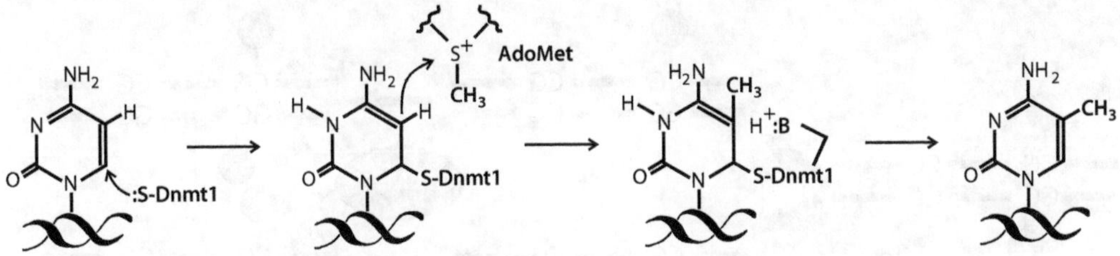

Fig. 1.7 Schematic of the catalytic mechanism of DNA methylation. DNMTs attach to the 6 position of the pyrimidine ring of the cytosine and enter a covalent intermediate. Methylation of the 5 position induces a shift of electrons and releases the enzyme

1.4.1 DNA Cytosine Methylation

The cytosine base of DNA in animal and plant genomes can be chemically modified to 5-methyl-cytosine (5mC). This modification is observed in the majority of animals but is conspicuously absent from some popular laboratory model organisms including the nematode *Caenorhabditis elegans*, the fly *Drosophila melanogaster*, and the yeasts *S. cerevisiae* and *Schizosaccharomyces pombe*. Although 5mC is not ubiquitous, it is widely distributed among animals[7] and plants. In particular, DNA methylation is essential for mammalian development and has been extensively studied for its role in silencing tumor suppressor genes in certain types of human cancer (see book ► Chap. 8 by Santoro).

5mC is catalyzed by the activity of DNA methyltransferases (DNMTs) that use S-adenosyl-methionine (SAM) as a methyl-donor. The catalytic center of mammalian DNMTs resembles the one of DNA methylases that are components of bacterial restriction systems. The reaction involves an attack at the 6 position of the cytosine ring by the thiol group of a cysteine, leading to the formation of a covalent bond between the cytosine and the DNMT. The reaction mechanism of DNMTs thereby comprises an intermediate that links DNMTs temporarily to DNA (■ Fig. 1.7). This intermediate is subsequently resolved by transfer of a methyl-group from SAM to the 5 position of the cytosine ring, abstraction of a proton, and release of the DNMT enzyme. After release, DNMT enzymes are available for another reaction cycle. Derivatization of the ring system of a DNA base requires that the base is accessible. From structural analysis of bacterial DNA methylases, it has been proposed that the cytosine is flipped out from the DNA double helix by rotation of the phospho-deoxyribose backbone. In this way, the base can be inserted into a deep pocket where SAM and catalytic residues are in close contact to the 5 and 6 positions of the cytosine ring.

A variety of methods has been developed to analyze DNA methylation at specific genes and genome-wide (► Method Box 1.2). 5mC in mammalian genomes is preferentially found in the context of CG dinucleotides, but not on cytosines in other sequence contexts like CC, CA, and CT. The reason for this becomes clear when replication of the DNA and the maintenance of methylation patterns are considered. CG dinucleotides represent a symmetric configuration on the antiparallel DNA strand as C pairs with G and DNA strands have opposite polarity in the double helix.

7 DNA methylation has been observed in honey bees and ants suggesting that it is present in insects. The DNA methylation system was likely lost during the evolution of *D. melanogaster*.

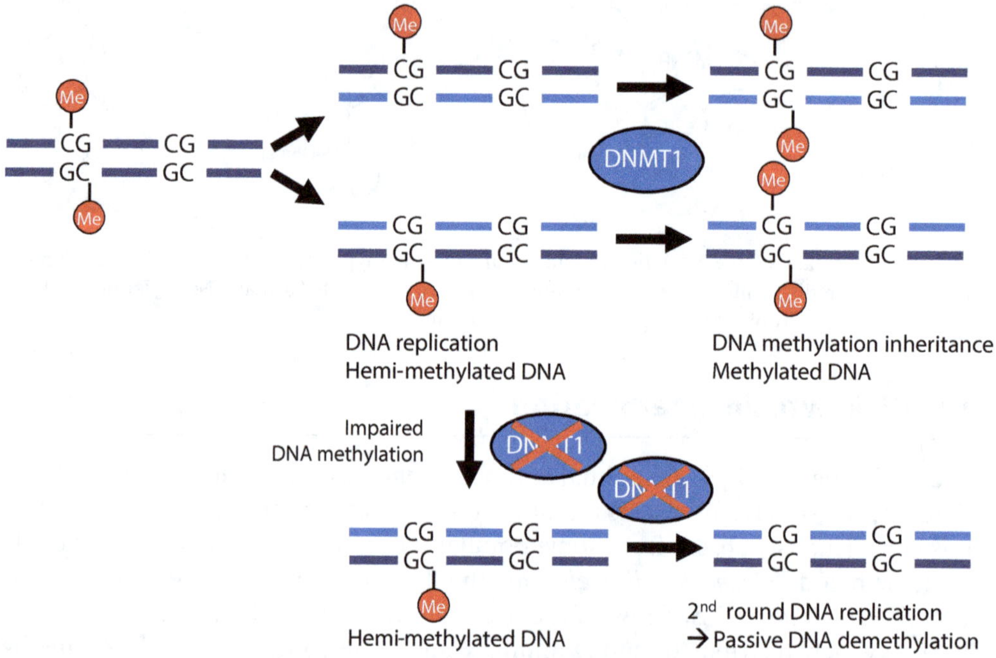

Fig. 1.8 A model for maintaining CpG methylation patterns in mammalian genomes. After a single round of DNA replication, hemi-methylated DNA is generated. Restoration of DNA methylation on the newly synthesized DNA strand leads to a heritable pattern of DNA methylation. DNMT1 is an enzyme that is targeted to hemi-methylated substrates and has maintenance DNMT activity. The original 5mC is not lost but unmethylated C is present in newly synthesized DNA. Hence, if CpG methylation patterns are not restored after 2 rounds of DNA replication, fully unmethylated (newly synthesized) DNA is expected (bottom). This leads to a loss of epigenetic information by passive, replication-dependent demethylation

If DNA with symmetrically methylated CG dinucleotides on both strands is replicated, two copies result, each having a methylated and an unmethylated (the newly synthesized) strand. Maintenance DNMT enzymes are recruited to hemi-methylated DNA to restore a fully symmetric methylation pattern (Fig. 1.8). This mechanism requires the recognition of hemi-methylated CG dinucleotides and the activation of methylation activity to restore the pattern on the newly synthesized strand for a faithful maintenance of the epigenetic marks. This conceptually simple mechanism can explain the heritability of this particular epigenetic information through cell division. We will see shortly that the molecular details are not quite as simple but the overall concept is clear and easy to comprehend.

Method Box 1.2: Analysis of DNA Modifications

DNA methylation can be detected by different methods. Bacterial restriction endonucleases that are inhibited by cytosine methylation can be used to digest genomic DNA. Electrophoretic size separation and hybridization with complementary probes (Southern blot) reveals unmethylated DNA at a lower molecular weight than the methylated DNA, which remains undigested. PCR with gene-specific primers can

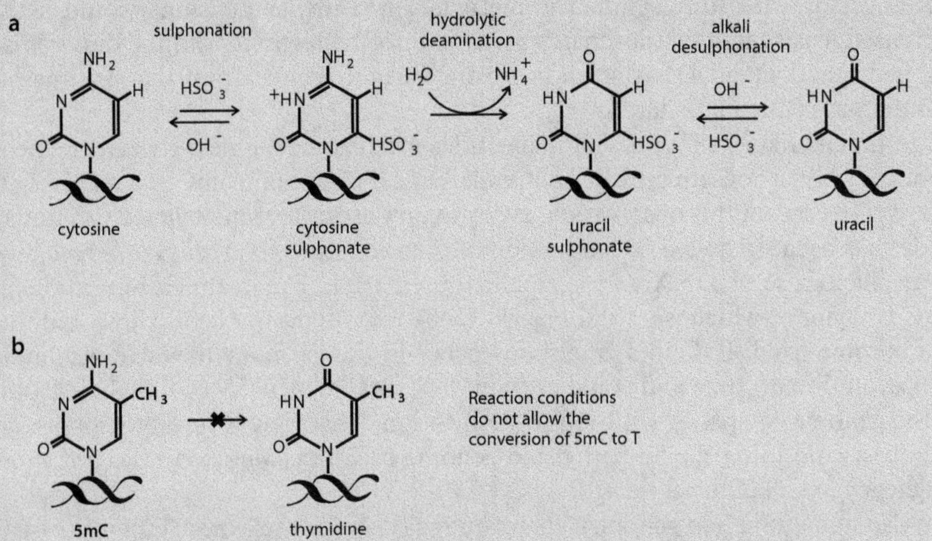

Box Fig. 1.2 Analysis of 5mC by bisulfite conversion. **a** The reaction mechanism for the conversion of cytosine to uracil in the presence of bisulfite. **b** The conversion of 5mC to thymine is blocked under suitably chosen reaction conditions

also be used to detect methylated DNA. When the endonuclease cuts the unmethylated template, no PCR product is obtained. Suitable controls are required as interpretation relies on the completeness of the enzymatic reaction. By using restriction enzymes that recognize the same sequence but are not inhibited by cytosine methylation, the digestibility of the DNA can be confirmed. *Hpa*II and *Msp*I recognize CCGG, but only *Hpa*II is inhibited by C5mCGG.

5mC can also be detected by antibodies. Similar to ChIP, methylated DNA is precipitated using 5mC-specific antibodies and can then be studied by a range of techniques including dot blots, PCR, and sequencing. Similar antisera have been developed for 5hmC. These antisera have further been used for immunofluorescence staining to observe the distribution of DNA modifications in tissues and cells by microscopy techniques.

Bisulfite conversion allows to examine DNA methylation at single base resolution. Heating DNA in the presence of bisulfite leads to deamination of cytosine to uracil (Box Fig. 1.2). Under suitable reaction conditions, the conversion of 5mC is not observed and, thus, cytosine remains unchanged. Sequencing of the converted DNA shows unmethylated cytosines as thymines, so that methylation can be identified by comparison to a reference sequence.

Recently, nanopore sequencing and mass spectrometry have also been applied to detect and quantify DNA modifications.

What would happen if methylation does not take place, possibly because the DNMT enzyme is inactive or unavailable? Well, after another round of cell division, genomes would emerge that have completely lost DNA methylation from both strands and, hence, the corresponding epigenetic information (Fig. 1.8).

We would expect this to happen to half of all cells that have descended from the last ancestor with fully symmetric methylation. Further divisions would lead to increased loss as methylation patterns would not be restored and the only remaining hemi-methylated DNA strands are the ones inherited from the last ancestral cell that was fully methylated.

In mammalian genomes, CG nucleotides are observed in less than the expected frequency.[8] As a consequence, long stretches of DNA contain 5mC at a relatively low density. To explain this observation, evolutionary erosion of methylated CG dinucleotides has been suggested. Deamination of cytosine leads to uracil that is recognized as an illicit base in DNA and is efficiently replaced. The deamination product of 5mC is thymine, which is a valid base in DNA and might not be repaired and, thus, can become fixed. If C to T mutations occur in the germline they can accumulate over evolutionary time and could explain the overall deficiency of CG dinucleotides in the genome of species with DNA methylation. This view is further supported by the observation that the unmethylated genome of *D. melanogaster* does not show a deficiency in CG dinucleotides.

Also in mammalian genomes, there are regions with the expected number of CG dinucleotides. These regions appear as islands of about 500–1000 bp of locally elevated CG density compared to the majority of the genome. These CpG islands[9] have been associated with gene promoters and are normally devoid of any 5mC. However, methylation of cytosines in CpG islands does occur in specific physiological contexts. CpG islands are methylated on the inactive X chromosome in female mammalian cells as a consequence of transcriptional repression (see book ▶ Chap. 4 of Wutz). In addition, methylation of CpG island promoters of tumor suppressor genes is important in human tumors (see book ▶ Chap. 8 of Santoro). It has been shown that loss of DNA methylation from the promoters of tumor suppressor genes can lead to their reactivation and cause cell cycle arrest and death of tumor cells. Therefore, DNMTs have been pursued as potential targets for treating human tumors (see book ▶ Chap. 8 by Santoro).

The physiological function of DNA methylation has been explored by reverse genetic analysis in mice. Mice possess three genes with catalytic DNMT activity. DNMT1 is considered the maintenance DNMT and catalyzes most of the replication-coupled cytosine methylation. This is the enzyme that we invoked in the simple mechanism for inheriting DNA methylation patterns earlier in this chapter. In addition, there are two enzymes, DNMT3A and DNMT3B, that are considered *de novo* DNMTs. These DNMTs are thought to newly establish DNA methylation patterns and are targeted by other factors to chromatin. All DNMTs possess characteristic sequence homology in their catalytic domains. In mice, DNMT1 and DNMT3B are essential for embryo development. Mutations of DNMT3B are also associated with the immunodeficiency, chromosomal instability, and facial abnormalities (ICF) syndrome in humans, suggesting a contribution of DNMT3 to centromere function and

8 In a stretch of mammalian DNA, fewer CGs occur than any of the other possible dinucleotide combinations GC, AT, TA, AC, CA, AG, GA, CT, TC, GT, TG, AA, CC, GG, TT. In plants, CNG appears the sequence motif instead of CG. Like CG, CNG is symmetric on both DNA strands.

9 The term CpG islands, the "p" representing the phosphor-ribose linkage, is often used instead of CG islands to indicate that CpG dinucleotides and not CG base pairs are meant.

Table 1.1 Components of the DNA modification systems that are mutated in human disease

Gene	Disease
DNMT3B	ICF syndrome
MeCP2	RETT syndrome
DNMT3A	Haematopoietic malignancies
TET2	Haematopoietic malignancies

genomic stability. Part of the defects caused by loss of DNMT1 are derepression of retrovirus-like genomic elements, so called intra-cisternal A-particles (IAPs). Loss of DNMT1 also causes cell death of differentiated cells but might be tolerated in transformed cells to some extent. Surprisingly, early embryonic cells appear to tolerate a loss of DNA methylation. Studies have shown that DNA methylation is critical for gene regulation. In particular, gene regulation by genomic imprinting is intimately linked to cytosine methylation (see book ► Chap. 5 of Grossniklaus).

Analyses of the effects of mutations on DNA methylation led to the identification of factors that are involved in the mammalian DNA methylation system. These include structural maintenance of chromosomes hinge domain 1 (SmcHD1), which contributes to DNA methylation and gene repression at promoters on the inactive X chromosome, the non-catalytic DNMT3L that acts together with DNMT3A in establishing DNA methylation patterns in the germline, and UHRF1 (also called NP95). UHRF1 is required for association of DNMT1 with the replication machinery. UHRF1 contains a domain with specificity for hemi-methylated DNA. Further studies have led to the discovery that UHRF1 possesses catalytic activity and mediates ubiquitinylation of histone H3 lysines 23 and 18. H3K23ub and/or H3K18ub recruit DNMT1 to DNA where it changes the hemi-methylated to a fully methylated pattern. It is interesting to note that histone modifications are an integral part of the maintenance pathway of DNA methylation patterns. We will see more overlap and crosstalk between DNA and histone modifications in the chapters throughout this book.

The methyl-group of 5mC is located in the major groove of the DNA double helix where it can be accessed by readers of DNA methylation. Protein domains have been identified that mediate specific binding to methylated or unmethylated DNA. UHRF1 contains such a domain for recognizing hemi-methylated DNA. A family of DNA-binding proteins contains a methyl-DNA binding (MDB) domain. In this family of proteins, methyl-cytosine binding protein 2 (MeCP2) specifically recognizes 5mC. The function of MeCP2 is not entirely clear but it appears to affect gene expression in a subtle way. Mutations affecting MeCP2 have been shown to cause RETT-syndrome in humans (◘ Table 1.1). RETT syndrome is named after its discoverer and characterizes a neurodevelopmental disorder. It mainly affects girls who initially appear to develop normally but, at about 1 year of age, regress and display neurologic symptoms that overlap with symptoms of autism. The mutation is in a heterozygous state and *MeCP2* is located on the X chromosome. Due to random

X inactivation, only about half of the patients' cells lack *MeCP2* expression whereas the other half is phenotypically normal. It has been suggested that the presence of MeCP2-deficient cells might have a dominant effect disrupting brain functions, which is likely due to subtle changes in gene expression in neurons and glia cells. A mouse model has been established that recapitulates some neurological defects to study the disease mechanism. Experiments that allow the restoration of *MeCP2* function after symptoms have developed, suggest that neurologic phenotypes can be reversed (Guy et al. 2007). This finding has spurred efforts to reactivate the intact copy of *MeCP2* in human patients where an intact copy of the gene resides on the inactive X chromosome. If these approaches were successful and *MeCP2* could be reactivated, possibly by removing DNA methylation, a treatment of this devastating disease could be found. However, caution is needed as the mouse model on which these hopes were based does not fully recapitulate all phenotypes that are observed in humans. Mice that lack MeCP2 completely are viable and show neurologic symptoms, whereas absence of MeCP2 in all cells is lethal in humans. Therefore, MeCP2 mutations are generally not observed in males.

A second way how DNA methylation can affect gene expression is by preventing the recognition of binding sites by transcription factors. One might think of a steric hindrance by methyl groups that stick out into the major grove of a DNA segment containing a binding motif. The transcription factor CCCTC-binding factor (CTCF) recognizes a motif containing cytosines that can be methylated. CTCF plays a role in chromatin organization and has been associated with the formation of chromatin boundaries and insulator function (see ► Sect. 1.7.3 of this Chap., and book ► Chap. 5 of Grossniklaus). CTCF binding is blocked by DNA methylation, which can have profound consequences for the genomic region.

Studies of DNA methylation in mouse development have revealed that early embryos lose most of their DNA methylation at the blastocyst stage (see book ► Chap. 5 of Grossniklaus). Initially, the genomes of the gametes are characterized by substantial methylation, which is removed during the cleavage stages of preimplantation mouse development. Both active and passive (through DNA replication) demethylation mechanisms have been considered. It is interesting to note that the precise mechanism of enzymatic removal of 5mC from DNA has not yet been established. It is thought that DNA repair pathways play a role as it appears impossible to chemically remove the methyl-group from 5mC without opening the pyrimidine ring system. This eraser mechanism is therefore different from mechanisms of erasing histone modifications and, thus, DNA methylation is considered a relatively stable epigenetic mark.

1.4.2 DNA Cytosine Hydroxymethylation

A fundamental advance in understanding DNA methylation in animal genomes has come from the discovery of a family of enzymes that convert 5mC in DNA. The ten eleven translocation (TET) family of proteins has three members in mice, TET1, TET2, and TET3. These are iron- and α-ketoglutarate-dependent dioxygenases that use molecular oxygen to oxidize the methyl-group of 5mC (◘ Fig. 1.9). The first

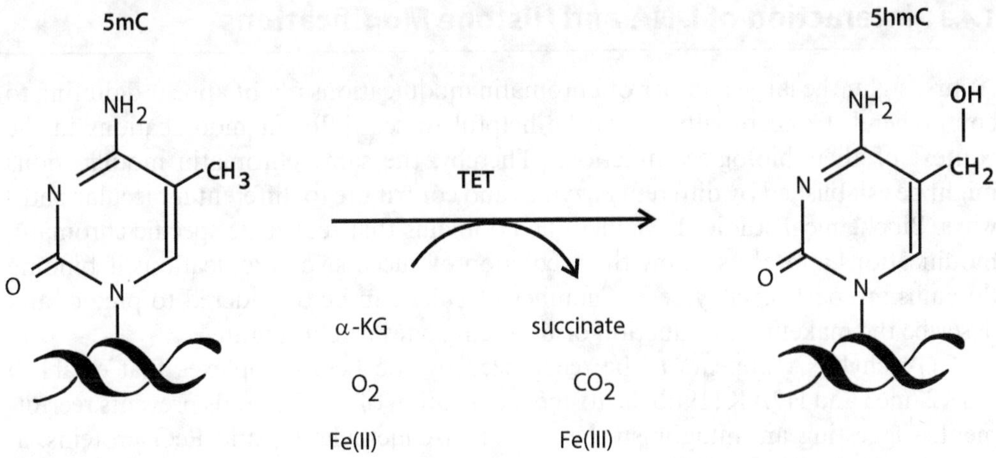

Fig. 1.9 Conversion of 5mC to 5hmC by TET dioxygenases. TET enzymes use oxygen to hydroxylate the methyl-group of 5mC. The reaction requires iron in oxidation state 2 [Fe(II)] and leads to the decarboxylation of α-ketoglutarate to succinate

product of this oxidation is 5-hydroxy-methyl-cytosine (5hmC). Further oxidation converts 5hmC to 5-formyl-cytosine (5fC), and 5-carboxy-cytosine (5caC). It has been suggested that recognition and removal of these oxidation products by the base excision DNA repair machinery could be a mechanism for active demethylation of DNA. Additional roles for different cytosine modifications have been proposed in gene regulation, organization of nucleosomes, and the regulation of histone modifications. It is important to recognize that the substrate for generating any of these modifications is 5mC and, therefore, DNMTs are required. The dependence of some modifications on a preceding reaction is interesting as a more complex regulation of epigenetic information might be established similar to an "AND" gate in Boolean logic. However, whether this aspect is used by cells is presently unclear. Also, constraints for the system of DNA modifications come from the observation that DNA methylation is dispensable in some organisms (*D. melanogaster* and *C. elegans*), suggesting that evolutionary important processes are not strongly dependent on it.

TET1 was initially discovered at the breakpoint of chromosomal translocations associated with leukemia and gave the family its name: ten-eleven-translocation. Mutation of *Tet2* in mice leads to aberrant blood cell differentiation and leukemia-like phenotypes (see book ► Chap. 8 of Santoro). These observations suggest a role of TET dioxygenases in the regulation of gene expression during cell differentiation. The function of *Tet3* has been associated with the demethylation of the genome in mouse preimplantation embryos. TET3 is present in the oocyte as a maternally supplied protein. After fertilization TET3 converts 5mC on the paternal, i.e., sperm-derived, genome to 5hmC. In contrast, the maternal genome is protected and maintains 5mC in the zygote. This leads to a differential marking of maternal and paternal chromosomes by 5mC and 5hmC, respectively. Although, the differences are removed through the subsequent passive demethylation process during embryonic cleavage, it is important to note that parent-of-origin marking is initially genome-wide. This is surprising considering that only a small number of imprinted genes maintain their parent-of-origin information throughout development.

1.4.3 Interaction of DNA and Histone Modifications

At first sight, the large number of chromatin modifications might appear daunting to comprehend. Conceptually, it can be helpful to see different modifications in the context of their biological functions. Thereby, the same chromatin modifications might be established by different enzymes and contribute to different molecular pathways. Biochemical studies have identified domains that recognize specific chromatin modifications and, thus, allow the prediction of successive modifications if binding domains are part of enzymes. A number of rules can be considered to predict and describe the makeup and function of different states of chromatin.

CG-rich DNA appears to be recognized by the PcG complexes that establish H3K27me3 and H2AK119ub. Cytosine methylation of CpG islands prevents recruitment, suggesting an antagonism between DNA methylation and PcG proteins, at least in CG-rich DNA. H3K27me3 has been shown to coexist with H3K4me3 that marks promoters. Such H3K27me3 and H3K4me3 doubly marked chromatin is considered to carry a "bivalent" mark, indicating that it has an ambivalent state between active and inactive. Yet, active promoters are acetylated on H3K27 and H3K27ac and H3K27me3 cannot coexist on the very same lysine. This logic suggests a potential step for activation by blocking PcG activity through H3K27ac (see book ► Chap. 3 of Paro). Notably, H3K9me3 does not appear to occur together with H3K4me3, such that a strong anti-correlation is observed. At pericentric heterochromatin, H3K9me3 is paired with H4K20me3 as we have seen earlier in this chapter. In addition, it is correlated with H3K4me1. Preventing methylation of histone H3 lysine 4 causes DNA *de novo* methylation. It has been shown that all mammalian DNMTs bind to H3K4 when it is unmodified. Binding leads to structural changes that activate *de novo* DNMT activity. In addition, DNMT3A and DNMT3B bind to H3K36me2 and H3K36me3 that are associated with the gene body of active genes. It is thought that DNA methylation within the transcription unit helps to prevent spurious transcription initiation, thereby reducing transcriptional noise. Undoubtedly, future studies have to extend our understanding of functional pathways in chromatin but keeping a few general rules in mind is helpful to evaluate epigenomic data and to decipher its biological relevance.

1.5 Chromatin Organization and Compartmentalization in the Cell Nucleus

We have now taken a look at chromatin and identified molecular characteristics that can carry epigenetic information along the DNA sequence. This has led to the view of a highly functional structure of chromatin. We end this chapter with a brief analysis of how chromatin is arranged in the cell nucleus. This aspect becomes important when one considers the large number of nucleosomes encompassing an animal genome. In the case of the diploid human genome, over 40 million nucleosomes will need to be assembled. This has implications of how genes can be identified and how transcription factors can bind to their recognition sites. Transport is not the only issue. The RNA polymerase II complex that transcribes most messenger RNAs exerts force on the DNA fiber. Transcribing in the order of 10,000 genes in a cell would lead to excessive movement within the nucleus. Observations from microscopy suggest

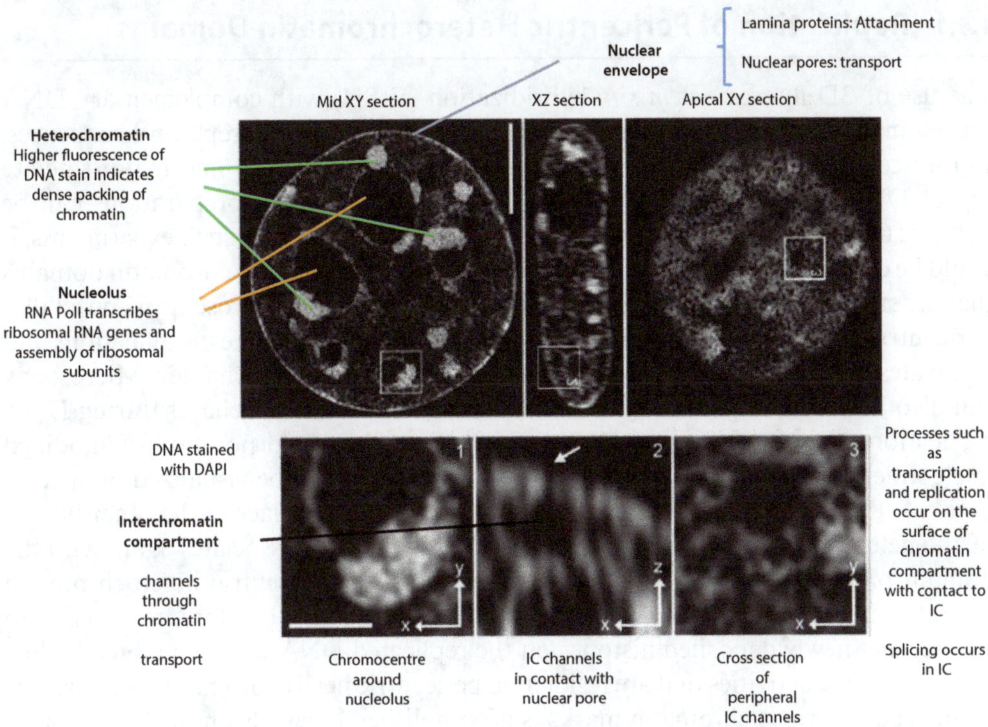

◘ Fig. 1.10 Organization of the cell nucleus. Super-resolution microscopy images of the cell nucleus have been annotated with prominent structures and compartments. (Adapted from Markaki et al. (2012))

that chromatin maintains a relatively stable distribution. Discrete foci are observed from pericentric heterochromatin, with similar chromatin appearing to cluster together. Also, chromosomes appear to occupy discrete volumes within the cell nucleus, referred to as chromosome territories (CTs). Although different CTs are in close contact, there is very little mixing of the majority of the DNA from different chromosomes. These observations constitute evidence that the chromatin fiber is constrained and mechanisms exist that compartmentalize the nucleus.

Microscopy images of DNA-stained nuclei show the nucleoli as prominent structures (◘ Fig. 1.10). These are the locations where ribosomal RNA is transcribed by RNA polymerase I and the assembly of ribosomal subunits takes place. Nucleoli contain RNA and ribosomes at different stages of assembly. As they contain little DNA, they stain weakly with DNA dyes and appear dark under the microscope. Nucleoli are bordered by a rim of heterochromatin called perinucleolar heterochromatin. Throughout the nucleus, DNA staining is distributed with different brightness corresponding to nucleosome density. Notably, small holes appear throughout the DNA staining. These are channels within chromatin, which are thought to allow genes to be accessed by factors and transcripts to be transported (Markaki et al. 2012). Connections of these channels to nuclear pores have been observed in super-resolution images, illustrating how mRNAs might be exported from the interior of the nucleus. Channels and gaps within chromatin are referred to as interchromatin compartment (IC). Collectively, these observations suggest that processes might take place on the surface of chromatin with factor access and transport being dependent on the IC.

1.5.1 Replication of Pericentric Heterochromatin Domains

The use of 3D fluorescent *in situ* hybridization (FISH) with complementary DNA probes makes it possible to correlate nuclear sub-domains with repetitive sequences. Fluorescently labeled DNA probes can be used to detect minor and major satellite repeat DNA of the pericentric regions in fixed nuclei. These preparations can be further stained using antibodies for histone modifications. From such experiments, it could be concluded that the pericentric repeats are organized into chromatin domains that are stained homogenously by H3K9me3 antisera. Thus, microscopy showed that pericentric heterochromatin is indeed marked with certain histone modifications and associates with specific proteins as we discussed earlier in this chapter. Microscopy can also be useful to gain insights into how heterochromatin behaves during DNA replication. Newly synthesized DNA can be marked by incorporation of modified nucleotides such as bromo-deoxyuridine (BrdU), which can be visualized by specific antisera. Replication of pericentric repeats appears to take place at the outer border of the heterochromatin domain. The replicated DNA then associates again with the domain of heterochromatin. As we have seen earlier, pericentric heterochromatin recruits enzymes that catalyze histone modifications. It is, therefore, reasonable to predict that newly deposited histones on the replicated DNA will be targeted by histone modifying activities that are resident in pericentric heterochromatin. Restoration of the characteristic chromatin marks is accomplished during S-phase and chromosomes with faithfully marked pericentric repeats can then be segregated in mitosis. Therefore, self-association of pericentric heterochromatin can be regarded as a mechanistic basis for the heritability of histone modifications. Although this is a plausible model, it is likely that the situation is more complex. Firstly, pericentric repeats are also subject to DNA methylation and, therefore, additional interactions might have to be taken into account to understand the mechanism. Secondly, it is presently not clear how self-association of pericentric chromatin is mediated and how it can be segregated from neighboring chromatin. Could it be an intrinsic property of chromatin modifications? Are specific H3K9me3 binding proteins mediating self-association by cooperative binding mechanisms? Potentially, additional factors could be involved in this process (see book ► Chap. 3 of Paro). Anchoring of chromatin at non-chromatin structures has been proposed, including a nuclear matrix or nuclear scaffold. However, the components of these conceptually plausible structures remain to be revealed.

1.5.2 Topologically Associating Domains

Even in images from super-resolution microscopy, which can resolve details of 30–100 nm, a single voxel[10] represents between 30 and 100 nucleosomes. Microscopy analysis is, therefore, on a different scale than our molecular understanding of reg-

10 A voxel represents the smallest volume that can be resolved in a 3D image. It is defined in analogy to a pixel, which represents a dot of a 2D picture.

ulatory elements. A typical 1 kb long promoter would be assembled into 7 nucleosomes. To bridge this gap, new methods have been developed. Chromatin conformation capture techniques (3C to HiC) aim to identify DNA fragments that are in close proximity of each other in the nucleus but might be far apart in the linear DNA sequence. The method preserves DNA contacts by cross-linking chromatin when it is in its native state in the cell (▶ Method Box 1.3). Thereafter, chromatin is digested with restriction enzymes. Ligation of the ends of the cut DNA in very dilute conditions yields chimeric DNA fragments. DNA fragments that are frequently in close proximity in the nucleus have a higher probability to be cross-linked in the same chromatin particle and, hence, show an elevated number of chimeric fragments. Sequence determination of the ligation products and plotting the number of chimeric DNA fragments against their distance in the linear DNA sequence yields a contact frequency estimate. Using this approach, the *in vivo* 3D topology of chromosomes has been investigated. From these studies it has become apparent that chromosomes are made up of domains of interacting chromatin that are partitioned by boundaries between them. Intervals with high internal contact frequencies are termed topologically associating domains (TADs, ▶ Method Box 1.3). Although TADs might share regulatory elements and be co-regulated, it is important to realize that 3C and related techniques usually investigate a large number – several thousands to millions – of cells and, therefore, only averages are observed. It is not always clear whether, in a given cell at a given time, all contacts take place. 3C techniques have been very successful in identifying gene regulatory elements such as enhancers that are located at a distance from their promoters but show a high contact frequency. Chromatin conformation data is, therefore, useful to determine regulatory interactions and possibly to identify genes that are coregulated by certain *cis*-regulatory elements. Recently, chromatin conformation capture-based techniques have been extended to single cell analysis. This has further contributed to our understanding of chromatin organization. From a study that used haploid G1 phase cells to make unambiguous assignments of DNA sequence contacts, it has become clear that a general subnuclear arrangement is adopted in all cells (Stevens et al. 2017). A-compartments predominantly consist of sequences that are correlated with active chromatin marks. These can be distinguished from B-compartments that contain largely inactive chromatin and frequently locate to the nuclear periphery. The precise winding and geometry of the chromatin fiber is different in each cell. Therefore, chromatin in the nucleus is stochastically distributed at the smallest scale but becomes more deterministic as the scale increases. A potential mechanism for such a topology has been proposed as local self-organization of chromatin, whereby locally acting elements would affect chromatin structure in their proximity. The combined activity of many elements would then result in the overall observed topology.

Method Box 1.3: Chromatin Conformation Capture (■ Box Fig. 1.3)

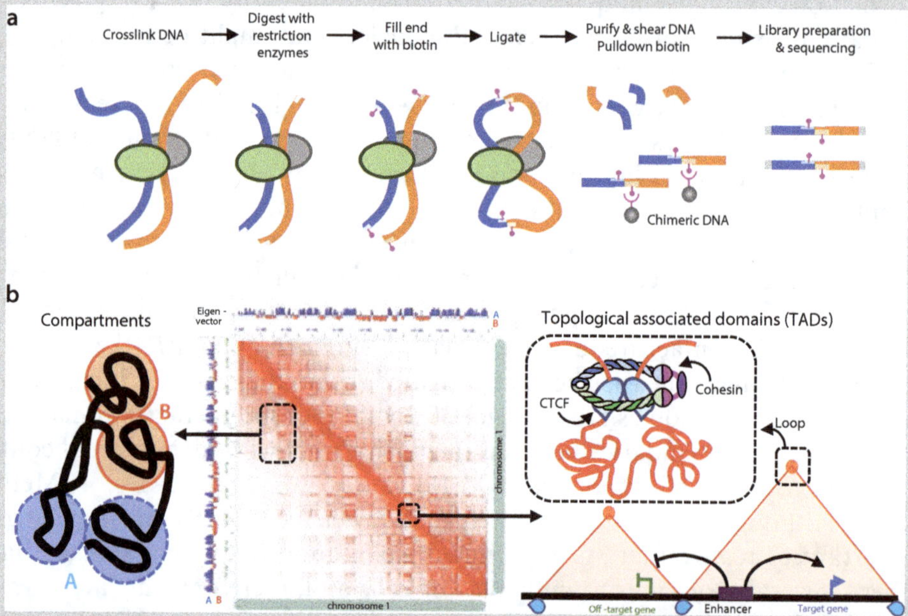

■ Box Fig. 1.3 Chromatin conformation capture. **a** The 3D organization of chromatin is preserved by crosslinking protein complexes and DNA with formaldehyde. The DNA is then digested with nucleases into short fragments. Subsequent ligation of DNA ends from fragments that are crosslinked facilitates the capture of interactions between sequences that are distant in the linear sequence of the genome but have been in close proximity in the nucleus. Sequencing of the ligation products and analysis of the frequency of chimeric DNA can be used to calculate the frequency of interaction between distant genomic regions. **b** Organization of chromatin compartments in the nucleus (left) can be deduced from interaction frequencies. A matrix of color-coded interaction frequencies (middle) can visualize the interaction of sequences in the genome. The diagonal line that runs from top left to right bottom represents interactions of regions that are in close proximity in the DNA sequence and are, thus, closely linked by the DNA backbone. Topological associated domains (TADs) (right) are confined by borders that limit interactions. High frequencies of interactions are observed within TADs and little interactions are observed outside. Cohesin complexes have been associated with TAD borders. Schematic representation of A and B compartments is also shown

1.5.3 Structural Maintenance of Chromosomes Complexes

The association of enhancers with gene promoters and the organization of TADs raises the expectation for mechanisms that facilitate the winding and bending of the chromatin fiber to achieve the observed topology. This aspect has been a long-standing question, which appeared difficult to answer for many years. Recent evidence illustrates a surprising and very powerful mechanism involving protein complexes of the structural maintenance of chromosomes (SMC) family. SMC proteins associate with chromatin and have important roles in chromosome condensation and segregation during mitosis. Two distinct SMC complexes have been identified, referred to as cohesin and condensin complexes in anticipation of their function in mediating sister chromatid cohesion and condensation of mitotic chromosomes, respectively. SMC proteins contain ATPase head domains that can hydrolyze ATP to exert motor function. Reconstituted recombinant condensin complexes (containing SMC1 and SMC3) have the ability to induce changes in the topology of fluorescently marked DNA molecules. This observation has been made in sophisticated microscopy-based biochemical assays (Ganji et al. 2018). DNA loops were formed when DNA fragments were incubated with condensin complexes and ATP. A similar function has subsequently been observed for a recombinant human cohesin complex (containing SMC2 and SMC4), when additional cohesin-associated factors were present (Kim et al. 2019). These experiments suggest the SMC family as important regulators of chromatin topology.

Cohesins have a role in the cell cycle where they mediate the cohesion of sister chromatids after replication of the chromosomal DNA until segregation at mitosis. In addition, cohesins have been found to be involved in gene regulation. Cohesins are loaded on chromatin in interphase and likely can slide along the chromatin fiber, mediating loop formation. Some chromatin loops correlate with the presence of CTCF sites at the loop base (▶ Method Box 1.3). CTCF binds sequences that are asymmetric and, therefore, have an orientation. CTCF sites at the base of chromatin loops are oriented towards each other, suggesting that CTCF might act in a directional manner to block cohesin sliding. A number of TAD boundaries can be correlated with CTCF binding sites. In addition, other molecules can also contribute to the localization of cohesins and possibly other SMC complexes. Therefore, the recent discovery of the DNA motor function of SMC complexes suggests a fundamental mechanism for establishing DNA topology at a larger scale that will certainly continue to attract attention in the future (Banigan and Mirny 2020).

Take-Home Message

- Chromatin organizes the genome of eukaryotic cells into a nucleosomal structure that facilitates the imposition of epigenetic information and assists in regulating the DNA sequence.
- The nucleosomal structure can be modified through histone variants, posttranslational histone modifications, and DNA modifications that act combinatorically and synergistically.
- Specific enzymes that establish histone modifications can be considered writers of epigenetic information. Conversely, readers are proteins with affinity for histone marks that can recruit protein complexes with additional functions to chromatin. Erasers are enzymes that remove chromatin marks and, thus, ensure that epigenetic information is reversible.
- Folding of the linear DNA in the nucleus facilitates the compartmentalization of chromatin into domains with different functional properties.
- Chromosomes are not randomly distributed but are organized into discrete volumes called chromosome territories, whereby separation of active A and repressed B compartments can be observed.
- Modifications of chromatin structure and chromatin topology both contribute to establishing information along the sequence of the DNA that regulates the function of distinct genomic regions.
- The understanding of nuclear pathways for chromatin regulation is a basis for the consideration of epigenetic mechanisms that will further be discussed in following chapters.

References

Banigan EJ, Mirny LA (2020) Loop extrusion: theory meets single-molecule experiments. Curr Opin Cell Biol 64:124–138. https://doi.org/10.1016/j.ceb.2020.04.011

Ganji M, Shaltiel IA, Bisht S, Kim E, Kalichava A, Haering CH, Dekker C (2018) Real-time imaging of DNA loop extrusion by condensin. Science 360(6384):102–105. https://doi.org/10.1126/science.aar7831

Guttman M, Amit I, Garber M, French C, Lin MF, Feldser D, Huarte M, Zuk O, Carey BW, Cassady JP, Cabili MN, Jaenisch R, Mikkelsen TS, Jacks T, Hacohen N, Bernstein BE, Kellis M, Regev A, Rinn JL, Lander ES (2009) Chromatin signature reveals over a thousand highly conserved large non-coding RNAs in mammals. Nature 458(7235):223–227. https://doi.org/10.1038/nature07672

Guy J, Gan J, Selfridge J, Cobb S, Bird A (2007) Reversal of neurological defects in a mouse model of Rett syndrome. Science 315(5815):1143–1147. https://doi.org/10.1126/science.1138389

Hirota T, Lipp JJ, Toh BH, Peters JM (2005) Histone H3 serine 10 phosphorylation by Aurora B causes HP1 dissociation from heterochromatin. Nature 438(7071):1176–1180. https://doi.org/10.1038/nature04254

Kim Y, Shi Z, Zhang H, Finkelstein IJ, Yu H (2019) Human cohesin compacts DNA by loop extrusion. Science 366(6471):1345–1349. https://doi.org/10.1126/science.aaz4475

Luger K, Mader AW, Richmond RK, Sargent DF, Richmond TJ (1997) Crystal structure of the nucleosome core particle at 2.8 A resolution. Nature 389(6648):251–260. https://doi.org/10.1038/38444

Markaki Y, Smeets D, Fiedler S, Schmid VJ, Schermelleh L, Cremer T, Cremer M (2012) The potential of 3D-FISH and super-resolution structured illumination microscopy for studies of 3D nuclear architecture: 3D structured illumination microscopy of defined chromosomal structures visualized by 3D (immuno)-FISH opens new perspectives for studies of nuclear architecture. BioEssays 34(5):412–426. https://doi.org/10.1002/bies.201100176

Olins AL, Olins DE (1974) Spheroid chromatin units (v bodies). Science 183(4122):330–332. https://doi.org/10.1126/science.183.4122.330

Schotta G, Lachner M, Sarma K, Ebert A, Sengupta R, Reuter G, Reinberg D, Jenuwein T (2004) A silencing pathway to induce H3-K9 and H4-K20 trimethylation at constitutive heterochromatin. Genes Dev 18(11):1251–1262. https://doi.org/10.1101/gad.300704

Stevens TJ, Lando D, Basu S, Atkinson LP, Cao Y, Lee SF, Leeb M, Wohlfahrt KJ, Boucher W, O'Shaughnessy-Kirwan A, Cramard J, Faure AJ, Ralser M, Blanco E, Morey L, Sanso M, Palayret MGS, Lehner B, Di Croce L, Wutz A, Hendrich B, Klenerman D, Laue ED (2017) 3D structures of individual mammalian genomes studied by single-cell Hi-C. Nature 544(7648):59–64. https://doi.org/10.1038/nature21429

2

Dynamic Structural and Nuclear Organization of Chromatin

Contents

What You Will Learn in This Chapter

The nucleus of a eukaryotic cell is a very busy place. Not only during replication of the DNA, but at any time in the cell cycle specific enzymes need access to genetic information to process reactions such as transcription and DNA repair. Yet, the nucleosomal structure of chromatin is primarily inhibitory to these processes and needs to be resolved in a highly orchestrated manner to allow developmental, organismal, and cell type-specific nuclear activities. This chapter explains how nucleosomes organize and structure the genome by interacting with specific DNA sequences. Variants of canonical histones can change the stability of the nucleosomal structure and also provide additional epigenetic layers of information. Chromatin remodeling complexes work locally to alter the regular beads-on-a-string organization and provide access to transcription and other DNA processing factors. Conversely, factors like histone chaperones and highly precise templating and copying mechanisms are required for the reassembly of nucleosomes and reestablishment of the epigenetic landscape after passage of activities processing DNA sequence information. A very intricate molecular machinery ensures a highly dynamic yet heritable chromatin template.

2.1 Basic Nuclear Activities

The eukaryotic genome is packaged into a nucleosomal beads-on-a-string architecture (see book ► Chap. 1 of Wutz). This allows the long but thin DNA fiber to be protected, condensed, and organized in the cell nucleus. For example, tight packaging of the genome is used actively to suppress deleterious sequences to become mobilized, like transposable or retroviral elements. However, DNA acts as a template for a number of nuclear processes which require access to the genetic information and, thus, an open chromatin conformation. Some of the associated enzymes use, for example, the sequence of a DNA strand to accomplish their work. However, nucleosomal DNA is double-stranded requiring that the DNA fiber be freed from histone packaging before these nuclear activities can be carried out. A number of mechanisms evolved to counter the primarily repressive nature of the nucleosome, permitting flexible and responsive nuclear activities.

A prime example in this respect is the process of *transcription*. The eukaryotic cell has three RNA polymerases, which are distinguished by the RNA products they produce. Their mode of action is similar, though, requiring the sequence information provided by an open DNA strand to synthesize the complementary RNA molecule (Cramer 2019). Before this process can start, gene activation mechanisms ensure that particular transcription factors (TFs) can access their specific target sequences, further requiring a remodeling of the protective nucleosomal structure at their sites of action, like enhancers. Similarly, during the process of elongation, the RNA polymerase complex encounters a nucleosomal structure, which needs not only to be removed for the synthesis of the transcript but also reassembled in an epigenetically conform manner after passage. All these steps are in conflict with the primarily static interaction of DNA and histones and require a dynamic chromatin structure responding to external signals, enzymatic processing, and mechanical forces.

The process of *DNA replication* uses the activity of DNA polymerases templating on single-stranded DNA the production of an exact copy of the genetic material (Hammond et al. 2017). As such, sites of origins, where replication starts, have to be freed of nucleosomes to allow the factors to bind DNA and assemble the replication machinery. Nucleosomes are removed ahead of the advancing replication fork and reassembled behind the site of replication. In addition, histone modifications, DNA methylation, and other chromatin associated factors are reestablished to maintain the epigenetic landscape. Indeed, at the replication site, two distinct mechanisms of epigenetic inheritance are taking place, of which we do understand one very well. The beauty of the complementary DNA structure revealed immediately how this molecule can self-template its own copy. Conversely, we know much less about how heritable epigenetic information is duplicated and faithfully replaced after passage of the replication fork. While we understand comparatively well the duplication of DNA methylation patterns on newly synthesized DNA (see book ► Chap. 1 of Wutz), we still lack a similarly comprehensive understanding of how protein-encoded information, including histone marks and histone variants is maintained in a heritable manner.

Other nuclear processes like *DNA repair* or *recombination* require the alignment of DNA sequences and, hence, a local removal of nucleosomal structures. Dedicated processes ensure fast repair of open damaged sites and the search factors required to align complementary genome sections for sequence exchange (Papamichos-Chronakis and Peterson 2013; Misteli and Soutoglou 2009).

Still other nuclear processes like the addition of *telomere sequences* to the end of chromosomes, the generation of anchoring points for the *segregation of chromosomes*, or the silencing of transposable or repeat elements in *heterochromatic sequences* are dependent on specific chromatin structures for their functional performance. This shows that chromatin needs to be dynamically regulated to support nuclear processes during the cell cycle in a predictable and reliable manner. The dampening nature of nucleosomes has also a positive effect, however. Transcription factors have many more presumptive binding sites in the genome than are observed *in vivo*. In combination with highly processive enzymes like RNA polymerase II, this can result in spurious and unwanted genome transcription. Such "expression noise" is kept in check by the nucleosomal packaging and only regulated chromatin accessibility will allow the basic nuclear processes to trigger their activities at the specific and required sites in the genome.

2.2 Connecting Nucleosomes to DNA Sequence

As packaging entities for an entire genome, it is clear that histones cannot rely on a similar DNA sequence-specificity as transcription factors. Nevertheless, as a nucleosomal unit, histone proteins are intimately interacting with the DNA that bends around the histone core. This can only be achieved through distinctive chemical interactions between the nucleic acid and the histones. As such, nucleosome assembly is thermodynamically preferred for some sequences in comparison to others, suggesting that chromatin architecture is governed by the sequence of the genome (Parmar and Padinhateeri 2020; Hughes and Rando 2014; Struhl and Segal 2013). Certain sequence determinants, like particular dinucleotides, affect DNA bending and, in consequence, nucleosome formation. Optimal nucleosome formation occurs when bendable dinucleotides (AT and TA) occur on the face of the helical repeat (approximately every 10 bp) that directly interacts with histones (◘ Fig. 2.1).

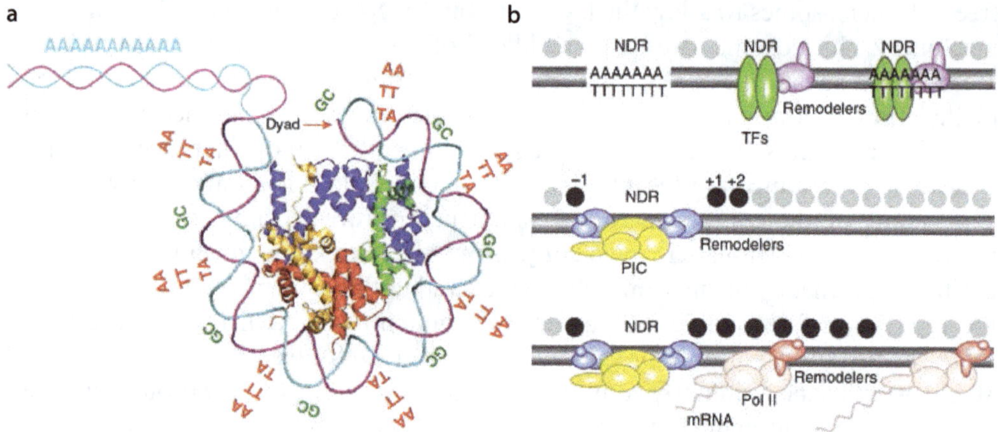

Fig. 2.1 Sequence determining nucleosomal position. **a** Nucleosome sequence preferences. Within the 147 bp that are wrapped around the histone octamer, there is a preference for distinctive dinucleotides that recur periodically at the DNA helical repeat and are known to facilitate the sharp bending of DNA around the nucleosome. These include ~10-bp periodic AA, TT, or TA dinucleotides that oscillate in phase with each other and out of phase with ~10-bp periodic GC dinucleotides. The linker regions exhibit a strong preference for sequences that resist DNA bending and, thus, disfavor nucleosome formation. Among these, poly(dA:dT) tracts and their variants are the most dominant and highly enriched in eukaryotic promoters. **b** Determinants of nucleosome positioning. (top) Nucleosome-depleted regions (NDRs) are generated either by poly(dA:dT) tracts and/or by transcription factors and the nucleosome remodeling complexes they recruit (see below). Gray circles indicate nucleosomes. (middle) Nucleosomes located at highly preferred positions (black circles) flanking the NDR are generated by nucleosome-remodeling complexes (for example, Isw2 and RSC, likely in a transcription-independent manner), and fine-tuned by the Pol II preinitiation complex (PIC) and associated factors. (bottom) Positioning of the more downstream nucleosomes depends on transcriptional elongation as well as the recruitment of nucleosome-remodeling activities (for example, Chd1 and Isw1) and histone chaperones by the elongating Pol II machinery. (From Struhl and Segal (2013))

The exact position of the histone octamer with respect to the 10-bp helical repeat of the DNA double helix is termed 'rotational positioning' and, thus, sequence is a critical determinant for the assembly of the nucleosome. However, protein-DNA interactions are not only recognizing base readout but are also dependent on the local shape structure of the DNA fiber, which is determined by the local sequence of bases (Rohs et al. 2010). Local kinks in the DNA fiber or variable width of the minor groove influence the efficiency of nucleosome build-up. Conversely, some homopolymeric sequences like poly(dA:dT) and poly(dG:dC) are rather stiff and are strongly inhibitory to nucleosome formation. Such sequences are often found at positions where nucleosome presence is reduced to allow transcription factors or polymerases access to enhancers or promoters, respectively. Hence, a combination of positively and negatively acting DNA sequences has a strong impact on the positioning of nucleosomes in the genome. Indeed, this information allowed predicting fairly accurately the position of nucleosomes in the yeast genome by computational methods. Similar attempts for more complex genomes have failed, probably because other factors might strongly influence the assembly and positioning of nucleosomes (see below). This indicates that sequence-dependent effects might be less important for nucleosome positioning in higher organisms.

In addition to a local interdependence between DNA sequence and nucleosome position, a number of DNA regulatory elements with a strong impact on large-scale chromatin conformations have evolved. Highly repressed chromatin, like heterochromatin, is built by multiprotein complexes with a strong polymerization activity. They package the nucleosomal fiber over an extended chromosomal region into a condensed and regular arrangement (see book ▶ Chap. 1 of Wutz). Heterochromatic regions do not only show a local regular array of condensed nucleosomes but also a clustering in the 3D space of the nucleus. This highly repressive structure can potentially immerse nearby active genes. The propensity of heterochromatin to spread along the chromosome can be observed in position effect variegation (PEV). After chromosomal translocations, repressive heterochromatin can "spread" into regions with euchromatic genes inhibiting their normal regulated activity. Normally, the presence of regulatory elements, termed boundary elements or insulators, restrains this expansive behavior of heterochromatin (Valenzuela and Kamakaka 2006). In cases where these insulators are lost, a stochastic spreading of heterochromatin in neighboring euchromatic genes can be observed, resulting in a variable expression of these genes.

However, also at the local level, different chromatin domains and gene regulatory units need to be appropriately controlled such that they can exert their function not over the linear dimension of the DNA fiber but in the true 3D space of the nucleus. At complex genetic loci, such insulator elements regulate the tissue-specific and developmental-specific interaction of distant enhancer elements with their corresponding promoter region. Additionally, such elements prevent the spreading of repressive chromatin, like PcG-chromatin (see ▶ Chap. 3 of Paro), into neighboring active expression domains. Technically, this knowledge can be used by flanking a transgene with insulator elements generating a functionally independent domain protected from position effects at the insertion site of the transgene. Appropriate regulatory signaling can occur within the insulated-flanked domain as the insulators prevent accidental external chromatin interactions.

Factors like sequence-specific CTCF bound to insulators do not only organize the local chromatin interactions, but, in conjunction with other chromatin-associated elements like cohesins, are also involved in large-scale chromatin loop formation or coordination of topologically associated domains (TADs; see book ▶ Chap. 1 of Wutz)(Chen and Lei 2019; Pisignano et al. 2019). Hence, at many points in the genome the DNA sequence plays, through specific DNA-protein interactions, an important structuring role in organizing chromatin and, consequently, contribute to the epigenetic landscape. Since every cell of a defined organism contains the same DNA, the question is how this apparent rigid structure of chromatin, nevertheless, permits a highly dynamic environment for the regulation of basic nuclear activities in a cell-type specific manner. In the next section of this Chapter, we will discuss how nucleosome positioning is regulated.

2.3 Nucleosome Remodeling

2.3.1 A Template for Transcription

This nuclear process transcribes RNAs from genes that are located in different regions of the genomic templates. These transcription units range from ribosomal RNAs, tRNAs, and mRNAs to the newly discovered miRNAs and long non-coding RNAs that are involved in distinct biological processes. In all cases, the process of transcription is tightly controlled in a temporal and spatial manner. RNA polymerases have to pass through three phases: initiation, elongation, and termination (Cramer 2019). The initiation phase involves recognition of a promoter DNA sequence, opening of DNA strands, and synthesis of a short initial RNA oligomer (Cairns 2009). During the elongation phase, the polymerase uses the DNA template to synthesize the growing RNA chain in a processive manner. Finally, DNA and RNA are released during termination, and the polymerase can then be recycled and re-initiate transcription. This entire process is reflected in a characteristic arrangement of nucleosomes and accompanying histone modifications at the transcription unit. The promoter region is devoid of nucleosomes while the downstream region exerts regularly phased nucleosomes as a consequence of the repeated passage of the polymerase complex (◘ Fig. 2.1c, d).

Biochemically, the sequence of events can be ideally studied at an inducible gene, for example, a target gene activated by a nuclear hormone receptor (◘ Fig. 2.2). Under such conditions, the process can be started and analyzed in a synchronous manner by the addition of a hormone to a tissue culture. After hormone binding, the nuclear receptor dimerizes and enters the nucleus. With the help of accessory factors, it can bind to its DNA recognition site even in a nucleosomal context. To open chromatin structures, in a next step, remodeling factors are attracted (see book ► Chap. 1 of Wutz and below). The unwound DNA can be bound by general TFs and, consequently, the transcriptional preinitiation complex can be loaded onto the promoter.

In many instances the cofactor SAGA is a multifunctional complex that facilitates transcription initiation in the context of chromatin. The TFIID complex also acts as a cofactor that is involved in promoter recognition. Both SAGA and TFIID can deliver the TATA box-binding protein (TBP) to promoters and stimulate assembly of the preinitiation complex. Eukaryotic RNA polymerases require a number of general transcription factors for initiation. Assembly of the polymerase on promoters to form a pre-initiation complex, for example, requires the general transcription factors TFIIA, TFIIB, TFIID, TFIIE, TFIIF, and TFIIH. In the case of Pol II, TFs do not target the polymerase directly but, rather, through interactions with cofactors, like the multiprotein complex Mediator. After DNA opening, the DNA template strand transitions into the active center of the polymerase. The template strand then directs *de novo* RNA chain initiation without the use of a priming oligonucleotide. Extension of the RNA to a critical length eventually triggers disassembly of the pre-initiation complex, escape of the polymerase from the promoter, and formation of a stable elongation complex.

The Pol II elongation complex often pauses downstream of the transcription start site (Gaertner and Zeitlinger 2014; Adelman and Lis 2012). Paused Pol II can be stabilized by the elongation factors DSIF and NELF but may also terminate (◘ Fig. 2.3). Upon an additional positive signal, Pol II is released from pausing into active elongation by the kinase positive transcription elongation factor, (P-TEFb). P-TEFb phosphorylates Pol II, DRB sensitivity-inducing factor (DSIF), and negative elongation factor (NELF), and this leads to dissociation of NELF and recruitment of elongation factors that stimulate Pol II progression. A number of epigenetic processes control transcription directly at the site of paused Pol II. Silencing by the *Polycomb* group (PcG) proteins (see book ► Chap. 3 of Paro) is the result of an inhibition of elongation by repressory PcG complexes. Transcriptional elongation is coupled to the processing of nascent RNA. This coupling requires the so-called

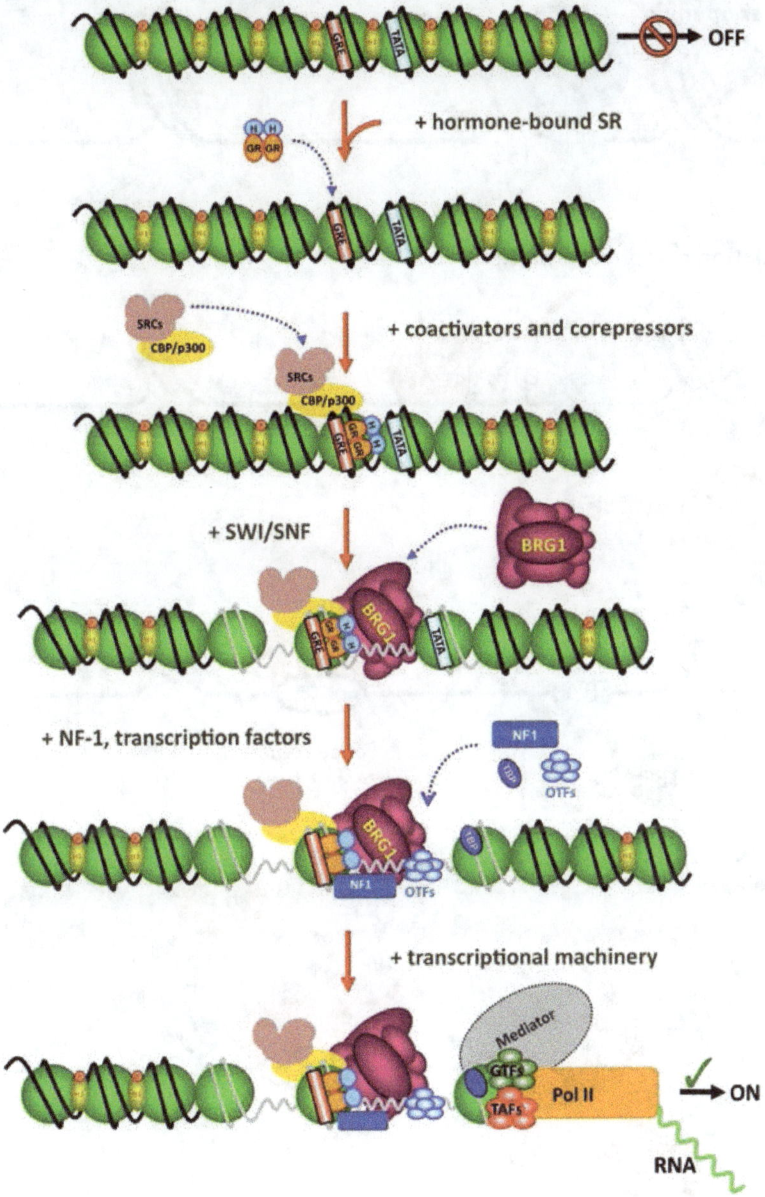

Fig. 2.2 Key transition points during hormone signaling. After ligand binding, glucocorticoid receptor (GR) dimerizes, enters the nucleus, and binds target sequences such as the Glucocorticoid Response Element (GRE) within chromatin. The receptor binds co-regulators, such as NcoA1/SRC-1, NCoA2/TIF-2/GRP-1, and CBP/p300, which do not require remodeling to engage the receptor at the promoter. In the following step, these factors attract directly the BRG1 nucleosome remodeling complex SWI/SNF to the promoter leading to the formation of a nucleosome free region. This allows transcription factors such as NF1 and the octamer transcription factors (OTFs) as well as the TATA-binding protein (TBP) to engage with the promoter. Finally, mediator proteins are recruited and the preinitiation complex forms, leading to transcription by RNA Pol II. (From King et al. (2012))

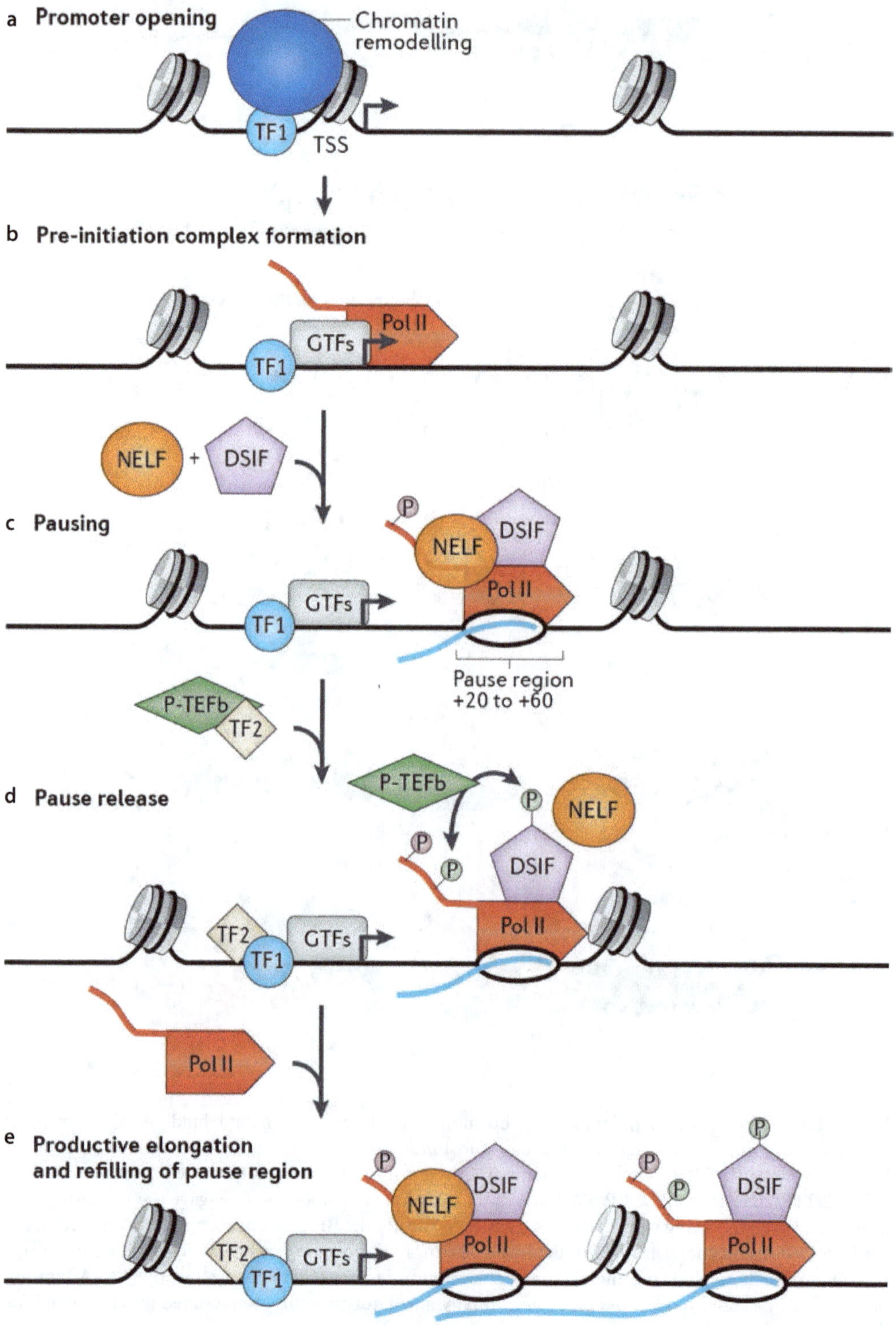
a Promoter opening
Chromatin remodelling
TF1
TSS
b Pre-initiation complex formation
Pol II
GTFs
TF1
NELF
+
DSIF
c Pausing
P
NELF
DSIF
Pol II
GTFs
TF1
Pause region +20 to +60
P-TEFb
TF2
d Pause release
P-TEFb
P
NELF
P
P
DSIF
Pol II
TF2
TF1
GTFs
Pol II
e Productive elongation and refilling of pause region
P
NELF
DSIF
Pol II
TF2
TF1
GTFs
P
P
P
DSIF
Pol II

C-terminal repeat domain (CTD). During elongation, the CTD is highly phosphorylated and recruits factors for post-transcriptional processing, like pre-mRNA capping, splicing, and also enzymes that modify histones. Indeed, the chromatin landscape at a gene appears to have a profound impact on post-transcriptional events and influences for example the specificity of splicing events. At the end of the transcribed region, Pol II transcribes the poly-adenylation site, which results in cleavage of the nascent RNA transcript and maturation of the mRNA.

At every point of the transcription cycle, the factors described above interact with a chromatin template and, therefore, require additional activities of partners to make DNA more accessible. Pioneer TFs can recognize their cognate DNA binding sequence in a nucleosomal context and open up chromatin by attracting histone acetylase complexes like SAGA that locally destabilize repressive nucleosomal structures. However, to make DNA accessible over an extended region, as required for example for the assembly of the large transcriptional pre-initiation complex, an energy dependent machinery in the form of ATP-dependent chromatin remodeling complexes is necessary.

2.3.2 Chromatin Remodeling Complexes

The process of remodeling involves reconfiguring the histone-DNA interactions either by disrupting, reassembling, or moving nucleosomes (Sundaramoorthy 2019). Enzymes known as ATP-dependent chromatin remodeling complexes are capable of generating changes by binding to a single nucleosome and reconfiguring it through utilizing the energy obtained from hydrolyzing ATP. Remodelers can (i) mediate nucleosome sliding in which the nucleosome position on DNA moves in either direction, (ii) generate access for TFs whereby histones remain bound but DNA becomes

◄

Fig. 2.3 Establishment and release of paused RNA polymerase II. The promoter region is shown with the transcription start site (TSS). Factors that are involved in the establishment or release of paused Pol II, such as DRB sensitivity-inducing factor (DSIF; purple pentagon), negative elongation factor (NELF; orange oval), and positive transcription elongation factor b (P-TEFb; green diamond), are indicated. **a** Promoter opening often involves binding a sequence-specific transcription factor (shown here as TF1) that brings in chromatin remodelers to remove nucleosomes from around the TSS and to render the promoter accessible for recruitment of the transcription machinery. **b** Pre-initiation complex formation involves the recruitment of a set of general transcription factors (GTFs) and Pol II, which is also facilitated by binding specific transcription factors (also shown as TF1 for simplicity). This step precedes the initiation of RNA synthesis. **c** Pol II pausing occurs shortly after transcription initiation and involves the association of pausing factors DRB sensitivity-inducing factor (DSIF) and negative elongation factor (NELF). The paused Pol II is phosphorylated on its carboxy-terminal heptapeptide repeat domain (CTD). The region in which pausing takes place is indicated in the figure. **d** Pause release is triggered by the recruitment of the positive transcription elongation factor b (P-TEFb kinase), either directly or indirectly by a transcription factor (shown here as TF2). P-TEFb kinase phosphorylates the DSIF–NELF complex to release paused Pol II and also targets the CTD. Phosphorylation of DSIF–NELF dissociates NELF from the elongation complex and transforms DSIF into a positive elongation factor that associates with Pol II throughout the gene. **e** In the presence of both TF1 and TF2, escape of the paused Pol II into productive elongation is rapidly followed by entry of another Pol II into the pause site, allowing for efficient RNA production. When the gene is activated, some nucleosome disruption is likely, as depicted by the lighter coloring of the downstream nucleosome. (From Adelman and Lis (2012))

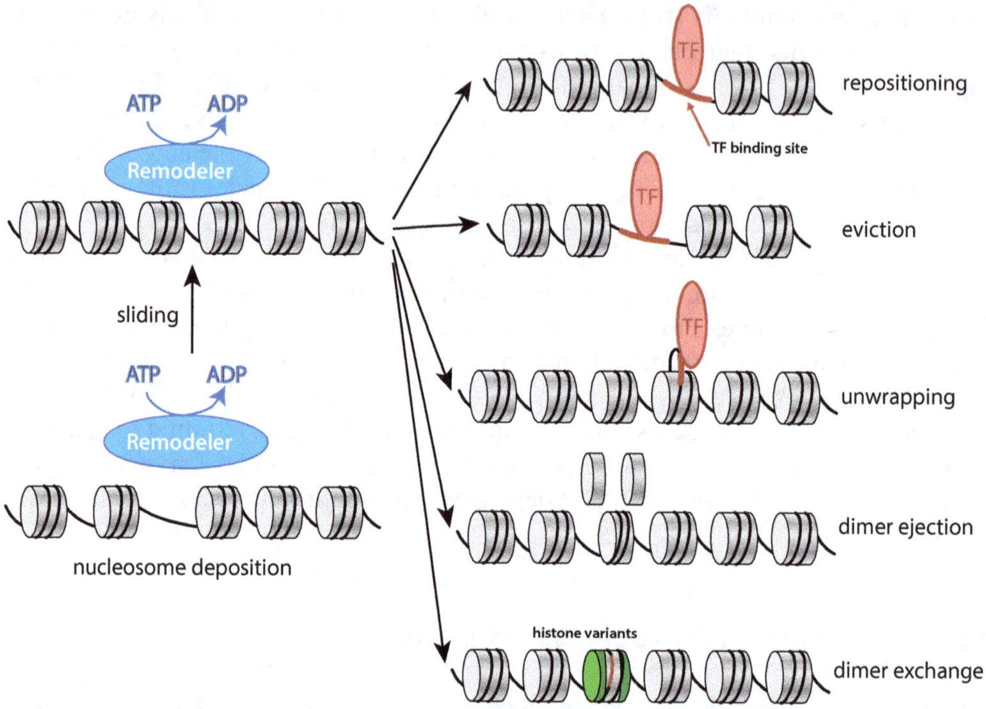

Fig. 2.4 The activities of chromatin remodeling complexes. Remodelers support assembly of a regular chromatin template by sliding deposited nucleosome and thereby providing space for new nucleosomes. Remodelers can make a binding site for a TF accessible by moving nucleosomes to the side (repositioning), by ejecting a complete nucleosome (eviction) or by partially freeing the DNA (unwrapping). In addition, remodelers can change the local composition of nucleosomes by ejecting a dimer or replacing histone dimers with corresponding variant forms (dimer exchange)

more accessible, (iii) induce eviction of the histone octamer, and (iv) replace a core histone with a variant histone (Fig. 2.4) (Lorch and Kornberg 2017; Clapier and Cairns 2009). The general importance of these processes is also evident by the fact that components of chromatin remodeling complexes are found to be frequently mutated in cancers (see book ▸ Chap. 8 of Santoro).

A shared property of this wide variety of effects rests on a common enzymatic function; an ATP-dependent DNA translocase activity that is utilized to break the histone-DNA contacts. This activity is carried out by an evolutionarily conserved catalytic core ATPase subunit. This remodeler ATPase subunit adopts a fold typical of the much larger superfamily 2 (SF2) DNA/RNA helicases and translocases. Analogous to SF2 family DNA translocases, the remodeler ATPase contains several highly conserved motifs, including WALKER A/B motifs, a DNA-binding cleft, and a site of ATP-binding that are all important for the DNA translocation activity. The ATPase of the remodeler differs from the SF2 DNA translocases in that they have much higher affinity for nucleosomes compared to naked DNA. An interesting aspect that emerged from the recently resolved CryoEM structures of various remodeler-nucleosome complexes is that remodelers unwrap outer turns of DNA from the nucleosome. This is a key step to minimize the number of histone-DNA contacts and maximize the output from ATP hydrolysis (Fig. 2.5). Based on the shared domains/motifs within their ATPase subunit fine-tuning the remodeler

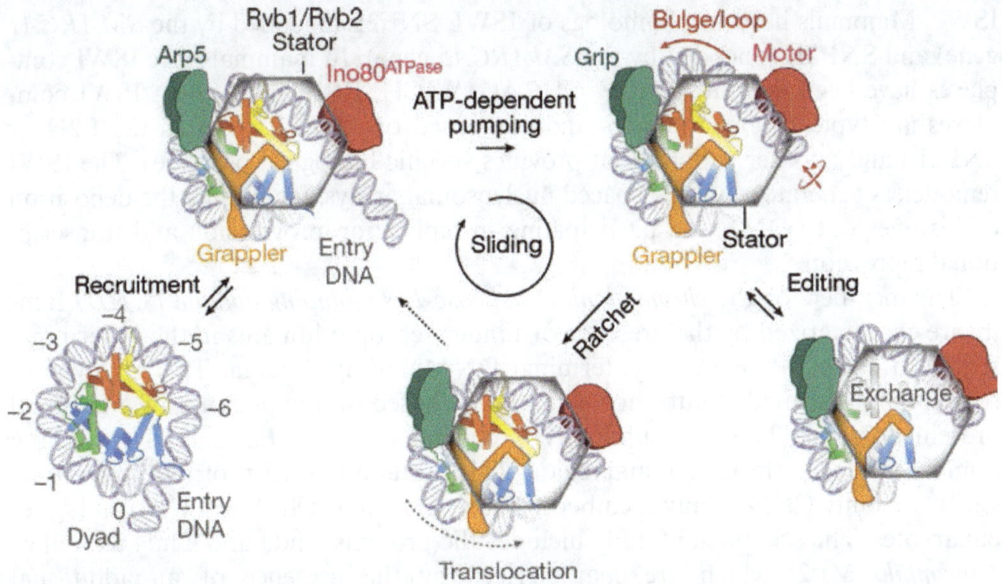

■ **Fig. 2.5** Model of INO80 nucleosome remodeling. The functional architecture of INO80 with motor (ATPase, red), grip (Arp complex, green) and grappler (orange) suggests that processive nucleosome sliding proceeds via a ratchet mechanism. Transient generation of loops between the motor and the grip could expose H2A–H2B for editing. Direct binding of H2A–H2B by the grappler sensorfoot could regulate variant- or modification-specific editing. (From Eustermann et al. (2018))

ATPase function and defining its selectivity for nucleosomal substrates, the family of Snf2 ATPases is grouped into four different classes: SWI/SNF, ISWI, CHD/Mi-2, and INO80-SWR1.

SWI/SNF remodelers are regulators involved in a wide range of gene control processes in all cell types. Composed typically of 9–14 subunits, at least two forms of SWI/SNF-specific Snf2 ATPases are present in most cell types with well-conserved signature domains: an acetyl lysine-binding bromo domain, a helicase-related SANT domain that binds actin-related protein (ARP), and the SF2 family helicase domain. In *Saccharomyces cerevisiae*, these two forms are termed as SWI-SNF (contains ATPase Swi2, also called Snf2) and RSC (contains ATPase Sth1). SWI/SNF complexes can oppose epigenetic silencing by PcG proteins (see book ▶ Chap. 3 of Paro). In *Drosophila melanogaster*, PcG proteins were shown to maintain repression of *Hox* genes during embryogenesis, while the SWI/SNF complex promotes *Hox* gene activation. In mammalian cells, the SWI/SNF complexes are generally grouped into two types of complexes: hBAF complexes containing ATPase BRG1 or BRM and several alternate core subunits such as ARID1A or ARID1B and hPBAF (polybromo-associated BAF) complexes containing the BRG1 ATPase associated with factors such as PBRM1 and ARID2. Both are essential for cell survival and involved in a number of transcriptional processes.

The *ISWI (imitation switch) remodelers* were initially identified in *Drosophila* and then subsequently observed in other species like yeast, *Xenopus*, and human. In *Drosophila*, the complex exists in at least four forms: ACF, NURF, CHRAC, and RSF. All four of these complexes possess the same ATPase protein ISWI but contain different associated subunits, which are likely important for specialized remodeler functions. In yeast *S. cerevisiae*, there are two distinct ISWI complexes, ISW1 and

ISW2. Mammals have two homologs of ISWI, SNF2L (encoded by the *SMARCA1* gene) and SNF2H (encoded by the *SMARCA5* gene). In mammals, five ISWI complexes have been identified: ACF, CHRAC, WICH, RSF, and NoRC. ISWI complexes are typically heterodimers and composed of an ISWI protein (SNF2H or SNF2L) and another subunit that provides specific biological processes. The ISWI remodelers generate regularly spaced nucleosomal arrays, facilitating the deposition of histones onto DNA and participating in replication progression and transcriptional repression.

The members of the *chromodomain-helicase-DNA-binding domain (CHD)* family are characterized by the presence of tandem chromodomains in the N-terminal region of the ATPase and a C-terminal DNA-binding domain. This highly conserved family is divided into three subfamilies based on the presence of additional protein domains. The first subfamily is defined by an additional DNA-binding domain located in the C-terminal region. This subfamily consists of the *S. cerevisiae* Chd1, the only CHD family member in yeast, as well as Chd1 and Chd2 in higher eukaryotes. The second subfamily includes the proteins Chd3 and Chd4 as well as *Drosophila* Mi-2, which are characterized by the presence of an additional N-terminal PHD domain. Members of the second subfamily are typically part of large multi-subunit complexes having besides chromatin remodeling functions, also histone deacetylation activities. The third subfamily contains the proteins Chd5–Chd9 and is characterized by additional functional motives in the C-terminal region, including paired BRK domains, a SANT-like domain, chromo domains, and a DNA-binding domain.

The remodelers of the *INO80 family* are the most recently identified remodeling complexes. The multi-subunit INO80 (Inositol 80) and SWR1 (Sick with Rat8) are the subfamily members of the INO80 family. INO80 regulates gene expression, DNA repair and replication by sliding nucleosomes, the exchange of histone H2A.Z with H2A, and the positioning of +1 and −1 nucleosomes at promoter DNA. Recent cryo-EM studies resolved the interaction sites of INO80 with a nucleosome (see ◘ Fig. 2.5) (Eustermann et al. 2018). Their analyses suggest a ratchet-like mechanism for how INO80 slides and possibly edits nucleosomes. INO80 unwraps and grips DNA and histones by multivalent interactions. The motor is positioned to pump DNA into the nucleosome against the Arp complex, which could hold onto DNA until a sufficient force is generated by multiple small steps of the motor. Such a tracking mechanism might create a DNA loop between the motor and the Arp complex. As a result, INO80 would move nucleosomes in larger steps of 10–20 bp. The proposed ratchet mechanism explains DNA loop formation that results in large translocation steps, as well as the means for ATP-dependent H2A.Z → H2A exchange. The motor, stator and multivalent grip of INO80 enable highly processive sliding without release and large-scale reconfigurations such as editing while keeping the remainder of the nucleosome intact.

Central to each of these remodeler families is the ATPase domain that conducts DNA translocase activity through ATP hydrolysis. Accessory domains within the ATPase and auxiliary subunits in a multi-subunit remodeler finetune the ATPase activity to the desired outcome of nucleosome remodeling. Hence, not only transcriptional processes require the moving capacity of the remodelers but all basic nuclear processes depend in one form or the other on this powerful competence to free DNA.

Methods Box 2.1: Determining DNA Accessibility in a Chromatin Template

As described in this chapter, the regulated access to the DNA template is a prerequisite for many nuclear enzymatic processes. In reverse, the appearance of accessible sites in chromatin could pinpoint to regions of regulation or active genomic processing. To this purpose, nuclei are treated with limited concentrations of nucleases like DNase I or micrococcal nuclease. Given the appropriate conditions, these enzymes will only cut exposed DNA but not DNA wrapped around histones or packaged in condensed or silenced nucleosomal structures like heterochromatin. Originally, open chromatin regions, termed DNase hypersensitive sites, could be determined at single genes. For example, after hormone induction, the chromatin rearrangements at the nuclear hormone binding sites and the accompanying changes at the promoter could be visualized by the appearance, over time, of DNase hypersensitive sites. Experimentally, after hormone induction of the cells in culture, the nuclei would be isolated and treated with nucleases under limiting conditions. The reaction would be stopped, the DNA isolated, its size further reduced with specific restriction enzymes, and eventually separated on an agarose gel. Hybridization with probes encompassing the boundaries of the gene would reveal the sites at which the nuclease attack had reduced the original size of the restriction fragment. Comparison of hypersensitive sites from uninduced and induced cells could provide not only information about the potential location of regulatory sites and produce co-called "DNase footprints" but also give clues about how fast chromatin is opening to permit the associated nuclear process (◘ Box **Fig. 2.1**).

In the meantime, recognition of hypersensitive sites can be achieved on a genome-wide scale. While the initial steps of limited nuclease attack to open chromatin conformations remains principally the same, the subsequent site recognition utilizes new high-throughput sequencing technologies. After nuclease digestion, fragments of different sizes appear; while nucleosome-packaged DNA, in general, produces larger fragments, regions with open chromatin or nucleosome depleted regions are cut more often, resulting in smaller fragments. Hence, a local high frequency of smaller fragments is an indication for a hypersensitive site in the genome.

A further improvement was achieved with ATAC-Seq (Assay for Transposase-Accessible Chromatin using Sequencing). Instead of DNase I, a mutated form of a transposase, Tn5, is used to fragment the chromatin. This enzyme not only cleaves DNA but also tags the resulting double-stranded DNA fragment with sequencing adaptors, which can be amplified by PCR and subjected to high-throughput sequencing. Also in this case, the number of reads for a region correlates with how open that chromatin region is. The method has the advantage that much less starting material, compared to the other described methods, is required. Overall, the method needs less time as fewer steps are required and tricky enzyme titrations are not necessary.

These methods have provided substantial insight into how chromatin "breathes" under specific physiological conditions, presenting a direct readout for chromatin dynamics. In all cases, however, the approach is dependent on nucleases which have an inherent albeit low DNA sequence specificity. This could result in preferential site cleavage, which is not solely governed by the chromatin structure. These caveats have to be considered when generating and interpreting results measuring chromatin dynamics.

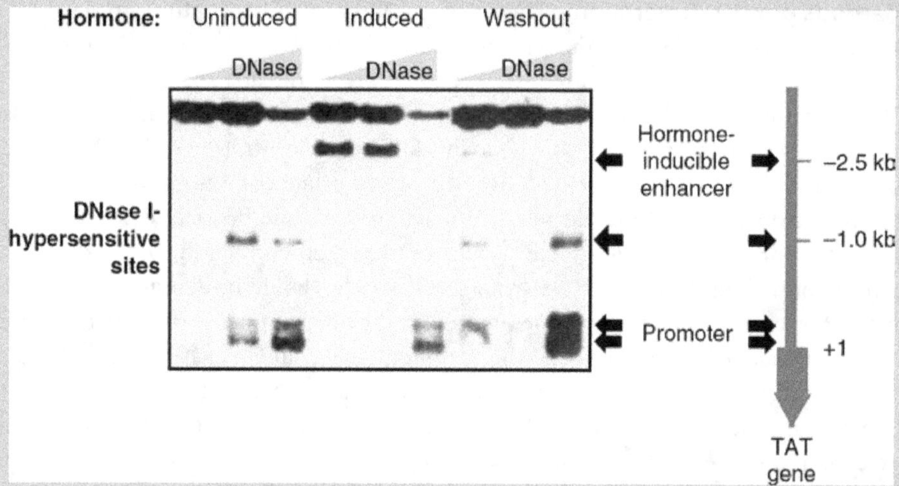

Box Fig. 2.1 DNase I-hypersensitivity (DH) analysis at an inducible gene. The chromatin organization at the glucocorticoid-responsive enhancer element, 2.5 kb upstream of the promoter of the tyrosine aminotransferase gene, was probed in rat liver cells. Isolated nuclei of cells are digested with increasing amounts of DNase I. Digested genomic DNA is purified, cleaved with a restriction enzyme, resolved by agarose gel electrophoresis, and subjected to Southern blotting. The DH sites are revealed by indirect end-labeling of restriction fragments through hybridization of a small radioactive probe (marked with arrows). In the uninduced state there are two DH sites at the promoter and one at 21 kb upstream. After hormone induction with corticosterone, nucleosomes are remodeled at the enhancer within 15 min. A new DH site appears 2.5 kb upstream of the transcriptional start site. This correlates with the binding of glucocorticoid receptors and a complex set of remodeling factors. On removal of the hormone ("washout"), the factors dissociate and canonical nucleosomes reform within 15 min and the 22.5-kb enhancer DH disappears. (From Becker and Workman (2013))

2.4 Nucleosome Assembly

The above section discussed processes evicting, moving, or replacing histones in nucleosomes and, thereby, liberating DNA for access to processing enzymes. However, the reverse process is as important in order to reconstitute the native nucleosomal template and the correct epigenetic landscape. This need is most evident at the time of DNA replication. The newly synthesized DNA has to be packaged and epigenetically marked in a regular and complementary manner. At the replication fork, a highly orchestrated mechanism not only templates and produces an identical copy of the DNA but also removes nucleosomes in front and reassembles histones into nucleosomes behind, now on two DNA fibers in an epigenetically impeccable state (Hammond et al. 2017).

2.4.1 Histone Variants and Histone Chaperones

Most histones are produced during the S-phase of the cell cycle. These DNA synthesis-coupled canonical histones provide the source for the repackaging of the two new chromatids. However, a number of DNA synthesis-independent histone variants are produced at any time in the cell cycle, with variable tasks in structuring the chromatin landscape (see also book ► Chap. 1 of Wutz) (Talbert and Henikoff 2017). The incorporation of the centromeric H3 variant (known as CENPA in vertebrates or CENH3 in plants) instead of the canonical H3 into a nucleosome forms the foundational structure of centromeric chromatin. The replacement of human canonical H3.1 by the highly similar variant H3.3 has few apparent consequences. This variant has a primary role in repairing gaps in the chromatin landscape that result from nucleosome disruption occurring during basic nuclear processes. Several H2A variants affect gene expression: H2A.Z and H2A.B are implicated in transcription initiation whereas macroH2A in animals and H2A.W in plants seem to be associated with nucleosome immobility and transcriptional silencing. Hence, like DNA containing base variants like methyl-cytosine, also nucleosomes can alter their information surface by incorporating variants of the canonical histones. This provides an additional layer of flexibility and complexity for epigenetic processing as the variants can have an influence on nucleosome stability and marking by chromatin modifiers.

For the assembly of histones into nucleosomes *histone chaperones* are required (De Koning et al. 2007). Indeed, when histones are mixed with DNA in solution at physiological ionic strength, they precipitate unless chaperones help histones to fold properly, preventing their positive charges from engaging in nonspecific interactions. The highly conserved histone chaperones are also necessary for escorting and depositing histones into chromatin and for histone storage. There are a number of different chaperones for cellular processes and pathways of histone assembly. Newly synthesized histones are bound by a series of chaperones in the cytosol before entering the nucleus. Once soluble histone complexes enter the nucleus, chromatin assembly factor 1 (CAF1) directs histone assembly behind the replication fork, whereas the histone regulator A (HIRA) complex directs assembly into chromatin independent of DNA synthesis (see below). In humans, CAF1 deposits canonical H3.1 and H3.2 during replication, whereas HIRA deposits the H3.3 variant throughout the cell cycle. Hence, any process in the cell requiring histone assembly utilizes the chaperoning capability of specific complexes. ◘ Figure 2.6 summarizes the identity and sites of action of these important factors in chromatin dynamics.

2.4.2 The Replication Fork: Still the Major Enigma in Epigenetics

The most outstanding cooperation between DNA, histones, histone chaperones, chromatin modifiers, and the replication machinery is exemplified at the replication fork (◘ Fig. 2.7) (Bellush and Whitehouse 2017; Hammond et al. 2017). The passage of genomic DNA through the small hole in the replicative helicase must result in the transient release of all DNA-binding proteins, including the disassembly of nucleosomes and, hence, the uncoupling of histone- and nucleosome-based epigenetic information from the DNA. Eventually, old nucleosomes, including many with histone variants and modifications, are rapidly reassembled on the newly synthesized leading and lagging strands. Gaps between old nucleosomes are filled with nucleosomes comprising newly synthesized canonical histones. Histone chaperones support this dynamic process. Behind the replication fork deposition of new nucleosomes is mediated by the histone chaperone CAF1, which teams with the replication clamp proliferating cell nuclear antigen (PCNA) and travels with the replisome. CAF1 is a trimeric complex

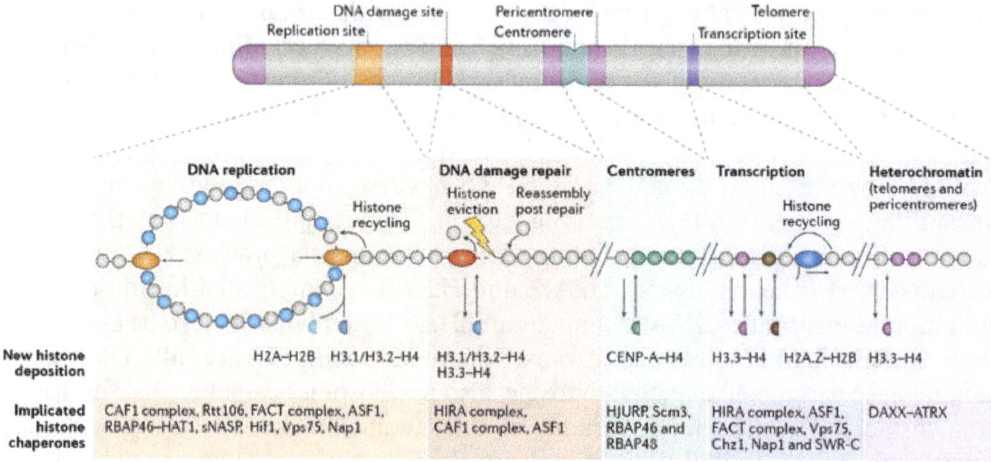

Fig. 2.6 Overview of histone deposition mechanisms. Newly synthesized histones are incorporated into chromatin via globally and locally acting mechanisms. A network of specialized histone chaperones controls histone delivery and deposition. The figure provides an overview of replication-coupled and replication-independent pathways that require the incorporation of newly synthesized canonical histones and replacement variants, together with parental histone recycling. The histone chaperones that are implicated in each process are listed at the bottom. (From (Hammond et al. (2017))

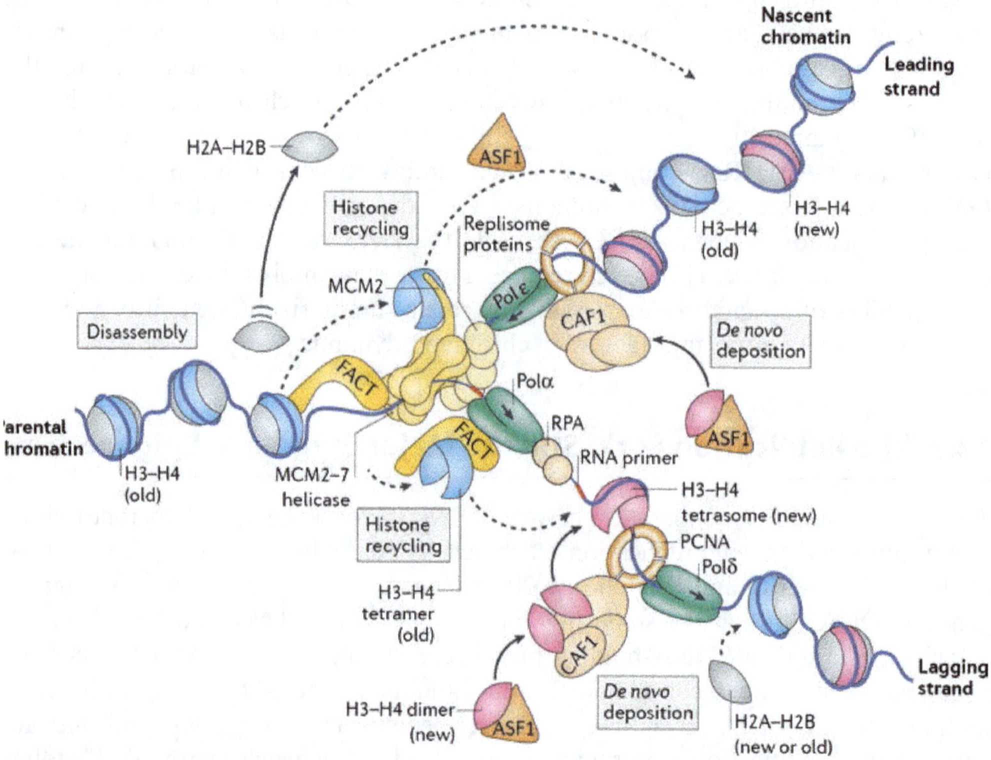

Fig. 2.7 Parental histone recycling during DNA replication. Evicted parental histones are randomly segregated to daughter DNA strands. It remains unknown whether other components of the DNA replication machinery have histone chaperone activity and whether deposition of old and new histones occurs by separate pathways (for details see text). (From Hammond et al. (2017))

responsible for assembling new canonical H3 nucleosomes behind the fork (see ◘ Fig. 2.7). ASF1 interacts with H3-H4 by binding the homodimerization interface of H3 and can only present dimers, not tetramers, to CAF1. A single CAF1 molecule can bind two H3-H4 dimers or a crosslinked $(H3\text{-}H4)_2$ tetramer, suggesting that CAF1 is capable of assembling and depositing $(H3\text{-}H4)_2$ tetramers onto DNA. Hence, old nucleosomes might be randomly disassembled and reassembled from their dimer components behind the fork, resulting in histone octamers containing both new and old histones. However, new Histones H3 and H4 may have epigenetic marks added before chromatin assembly at the fork. It is thought that H3K9 monomethylation and H4K5 and H4K12 acetylation are commonly added moieties by modifying enzymes SetDBl and HAT1, respectively, in complex with histone chaperones (e.g., CAF1 or ASF1). Once chromatin has been assembled these acetylation marks might become removed by deacetylases and replaced by corresponding opposing marks like H3K9me3 in the case of heterochromatin. However, in animals, canonical H3.1-containing nucleosomes consist of nearly all old or all new $(H3.1\text{-}H4)_2$ tetramers. This suggests that either old tetramers are transferred intact behind the fork or they are transiently disassembled and reassembled in a way that excludes the incorporation of new dimers. Transfer of tetramers or dimers occurs across the fork either to CAF1 via ASF1 or, perhaps with the assistance of FACT (Facilitates Chromatin Transcription), directly onto DNA. FACT may contribute to nucleosome disruption and histone recycling through interaction with the replicative helicase CMG complex (which includes CDC45-MCM2-7-GINS), polymerase α (Polα), and the single-strand binding replication protein A (RPA). MCM2 provides a binding platform for evicted H3-H4 tetramers and may facilitate their recycling directly, in collaboration with FACT, or as dimers with ASF1, which splits H3-H4 tetramers and forms a co-chaperone complex with MCM2.

The big question still is how, after the passage of the replication fork and the assembly of new and old nucleosomes, the local epigenetic landscape is reestablished (Singh 2014). A templating mechanism has to duplicate and appropriately pattern the new histones with the necessary modifications. Given the vicinity to the parental histones, these could act as anchors and templates for readers, writers and erasers to appropriately tag the new histones (see book ► Chap. 1 of Wutz). In light of the intricate patterns of modifications observed at certain functional chromatin locations, this certainly begs the question which and how precisely a biochemical mechanism is capable to faithfully reproduce these instructions. A better understanding of this process represents certainly a major and ambitious challenge in epigenetic research.

Take-Home Message

- The assembly of histones into nucleosomes is influenced by the DNA sequences wrapped around the protein core. Certain sequences favor bending of DNA whereas others disfavor the assembly of a nucleosome. Hence, the chromatin landscape is to a large part also determined by the genome sequence.
- DNA elements organize the chromatin in the 3D space of the nucleus. Insulator/boundary elements prevent long-range effects of repressive heterochromatin but can also locally control the interaction of *cis*-regulatory elements.

- DNA-templated processes like transcription are initiated by a well-controlled cascade of chromatin remodeling. Pioneer TFs provide sequence recognition without being inhibited by the nucleosomal structure and locally attract histone modifiers generating activating marks. ATP-driven chromatin remodeling complexes produce an extended local disruption of nucleosomes and partial replacement of canonical histones with variants. Correspondingly, open chromatin can be processed by polymerases to synthesize the required products.
- Either after DNA replication or after execution of other basic nuclear processes, the chromatin structure has to be fully reestablished. This does not only require the assembly of histones into a nucleosomal structure but also the maintenance of the appropriate epigenetic landscape defined by histone modifications, DNA methylation patterns, and others. The transport, storage, and local assembly of histones into nucleosomes is fostered by histone chaperones. In cooperation with the factors governing the basic nuclear processes and epigenetic processing, they ensure that the chromatin template is faithfully restored.

References

Adelman K, Lis JT (2012) Promoter-proximal pausing of RNA polymerase II: emerging roles in metazoans. Nat Rev Genet 13(10):720–731. https://doi.org/10.1038/nrg3293

Becker PB, Workman JL (2013) Nucleosome remodeling and epigenetics. Cold Spring Harb Perspect Biol 5(9). https://doi.org/10.1101/cshperspect.a017905

Bellush JM, Whitehouse I (2017) DNA replication through a chromatin environment. Philos Trans R Soc Lond Ser B Biol Sci 372(1731). https://doi.org/10.1098/rstb.2016.0287

Cairns BR (2009) The logic of chromatin architecture and remodelling at promoters. Nature 461(7261):193–198. https://doi.org/10.1038/nature08450

Chen D, Lei EP (2019) Function and regulation of chromatin insulators in dynamic genome organization. Curr Opin Cell Biol 58:61–68. https://doi.org/10.1016/j.ceb.2019.02.001

Clapier CR, Cairns BR (2009) The biology of chromatin remodeling complexes. Annu Rev Biochem 78:273–304. https://doi.org/10.1146/annurev.biochem.77.062706.153223

Cramer P (2019) Eukaryotic transcription turns 50. Cell 179(4):808–812. https://doi.org/10.1016/j.cell.2019.09.018

De Koning L, Corpet A, Haber JE, Almouzni G (2007) Histone chaperones: an escort network regulating histone traffic. Nat Struct Mol Biol 14(11):997–1007. https://doi.org/10.1038/nsmb1318

Eustermann S, Schall K, Kostrewa D, Lakomek K, Strauss M, Moldt M, Hopfner KP (2018) Structural basis for ATP-dependent chromatin remodelling by the INO80 complex. Nature 556(7701):386–390. https://doi.org/10.1038/s41586-018-0029-y

Gaertner B, Zeitlinger J (2014) RNA polymerase II pausing during development. Development 141(6):1179–1183. https://doi.org/10.1242/dev.088492

Hammond CM, Stromme CB, Huang H, Patel DJ, Groth A (2017) Histone chaperone networks shaping chromatin function. Nat Rev Mol Cell Biol 18(3):141–158. https://doi.org/10.1038/nrm.2016.159

Hughes AL, Rando OJ (2014) Mechanisms underlying nucleosome positioning in vivo. Annu Rev Biophys 43:41–63. https://doi.org/10.1146/annurev-biophys-051013-023114

King HA, Trotter KW, Archer TK (2012) Chromatin remodeling during glucocorticoid receptor regulated transactivation. Biochim Biophys Acta 1819(7):716–726. https://doi.org/10.1016/j.bbagrm.2012.02.019

Lorch Y, Kornberg RD (2017) Chromatin-remodeling for transcription. Q Rev Biophys 50:e5. https://doi.org/10.1017/S003358351700004X

Misteli T, Soutoglou E (2009) The emerging role of nuclear architecture in DNA repair and genome maintenance. Nat Rev Mol Cell Biol 10(4):243–254. https://doi.org/10.1038/nrm2651

Papamichos-Chronakis M, Peterson CL (2013) Chromatin and the genome integrity network. Nat Rev Genet 14(1):62–75. https://doi.org/10.1038/nrg3345

Parmar JJ, Padinhateeri R (2020) Nucleosome positioning and chromatin organization. Curr Opin Struct Biol 64:111–118. https://doi.org/10.1016/j.sbi.2020.06.021

Pisignano G, Pavlaki I, Murrell A (2019) Being in a loop: how long non-coding RNAs organise genome architecture. Essays Biochem 63(1):177–186. https://doi.org/10.1042/EBC20180057

Rohs R, Jin X, West SM, Joshi R, Honig B, Mann RS (2010) Origins of specificity in protein-DNA recognition. Annu Rev Biochem 79:233–269. https://doi.org/10.1146/annurev-biochem--060408-091030

Singh J (2014) Role of DNA replication in establishment and propagation of epigenetic states of chromatin. Semin Cell Dev Biol 30:131–143. https://doi.org/10.1016/j.semcdb.2014.04.015

Struhl K, Segal E (2013) Determinants of nucleosome positioning. Nat Struct Mol Biol 20:267–273

Sundaramoorthy R (2019) Nucleosome remodelling: structural insights into ATP-dependent remodelling enzymes. Essays Biochem 63(1):45–58. https://doi.org/10.1042/EBC20180059

Talbert PB, Henikoff S (2017) Histone variants on the move: substrates for chromatin dynamics. Nat Rev Mol Cell Biol 18(2):115–126. https://doi.org/10.1038/nrm.2016.148

Valenzuela L, Kamakaka RT (2006) Chromatin insulators. Annu Rev Genet 40:107–138. https://doi.org/10.1146/annurev.genet.39.073003.113546

3

A Comprehensive Study of Cellular Memory

Contents

What You Will Learn in This Chapter

The identity of cells in an organism is largely defined by their specific transcriptional profile. During cell division, these profiles need to be faithfully inherited to the daughter cells to ensure the maintenance of cell structure and function in a cell lineage. Here, you will learn how two groups of chromatin regulators, the *Polycomb* group (PcG) and the *Trithorax* group (TrxG), act in an antagonistic manner to maintain differential gene expression states. Members of the PcG cooperate in large multiprotein complexes to modify histones with repressive marks, resulting in condensed chromatin domains. Conversely, the TrxG proteins counteract the repressed domains by opening nucleosomal structures and establishing activating histone modifications. PcG and TrxG proteins are evolutionary highly conserved and control diverse processes, such as the identity of stem cells in mammalian development to the process of vernalization in plants.

3.1 Maintaining Cellular Fates

Metazoan organisms reach their size and body complexity through a sequential series of cellular specifications intermingled with phases of growth. During early embryonic stages, the body pattern is defined, the germ layers established, and the primordia for the organs delineated. Subsequently, during the process of differentiation the developmental fates are implemented to produce morphological features, organ structures, and functions fulfilling the needs of survival and reproduction of the specific organism. During all these development processes, cells also undergo a large number of divisions, with the requirement to retain the once established developmental fates[1]. To a large part, the fates and identities of cells are written in their differential patterns of gene expression. Hence, after each cell division, these unique transcriptional networks have to be reestablished to maintain cellular identities. This is not a trivial task given the extensive molecular and cellular rearrangements observed during the processes of genome replication and cellular division, where most regulatory proteins and chromatin factors are stripped from the DNA. However, complex mechanisms evolved to retain cellular identities in the wake of these processes, which ensure that developmental decisions are faithfully maintained. In other words, cells seem to have a *"cellular memory"*.

How can the "memory" of cells be revealed? Transplantation experiments have demonstrated that cells and tissues react differently when moved to a foreign neighborhood, depending on their programmed state. Before being determined for a particular identity, transplanted cells often adapt to the identity of their new foreign neighbors. Conversely, when transplanted after having been determined, they retain their original identity even when exposed to a new developmental environment. This shows that there is a specific period in the development of an organism in which cellular programs become fixed and irreversible (Wolpert et al. 2015).

1 Developmental fates are built through consecutive developmental restrictions. Among the first are the allocations to the germ layers (ecdoderm, endoderm and mesoderm). Subsequently, cells receive information to be part of a specific organ or tissue and fulfill a particular structural or functional role. Every step of these decisions have to be memorized and faithfully maintained.

In *Drosophila melanogaster*, at metamorphosis, larval cell are eliminated and adult precursor cells become differentiated during the subsequent pupal stage, giving rise to the organs and appendages of the adult fly. This process is triggered by the hormone 20-hydroxyecdysone (ecdysone). The burst of hormone induces a well-orchestrated cascade of transcriptional and post-transcriptional events leading to the differentiation of adult precursor (imaginal) cells. An example being the transformation of the morphologically inert imaginal disc cells which give rise to the highly specialized external cuticular structures, like wings, legs, or eyes. Concomitantly, larval cells and tissues which do not form part of the adult fly become marked for death and are histolyzed during metamorphosis. A classical experiment by Hadorn and collaborators demonstrated that imaginal discs maintain and remember their determined program (Hadorn 1968). Clusters of cells representing the primordia of imaginal discs are defined at early stages of embryogenesis. During subsequent embryonic and larval stages, theses clusters grow in size but do not differentiate. Only at metamorphosis, the cells implement their early defined transcriptional programs and differentiate into the structures and appendages that give rise to the external cuticle of the fly. Fragments of imaginal discs can be transplanted into the fluid of the abdomen of adult flies (hemolymph), which acts like an incubator, allowing the discs to grow in size but not to differentiate. This process can be repeated over many weeks and months, forcing the imaginal disc cells to undergo more cell divisions than they would have done during normal development of a fly. To observe whether the cells also retained their identity under these circumstances, they are transplanted back into a larva prior to metamorphosis. After undergoing metamorphosis, the hatched adult fly contains, in its hemolymph, differentiated cuticular tissue of the same identity as the original imaginal disc. This clearly demonstrates that cells can retain a memory of their programmed state even when placed in a foreign environment and undergoing many cell divisions.

Interestingly, Hadorn and collaborators observed rare cases where the transplanted imaginal disc fragments, after differentiation, produced cuticular structures other than those originally determined. Hence, cells can also "forget" and change fate. They termed this phenomenon *"transdetermination"*. These changes of fate often mimicked phenotypes which had been observed in homeotic mutants. These are mutations in master regulatory genes required to establish the appropriate segmental identities in the fly (Lewis 1978). As such, it was argued that the maintenance of cellular memory is inherently connected to the faithful conservation of the specific expression patterns of master developmental regulators.

3.2 PcG/TrxG System Maintaining Cellular Memory

Parallel to the studies of Hadorn, geneticists identified fly mutations displaying multiple homeotic phenotypes. Among these, Ed Lewis and his wife Pam characterized a mutation they named *"Polycomb"*, which they demonstrated to control the homeotic master regulators in the Bithorax Complex (BX-C) (Lewis 1978). In this mutant, the cuticular structures of the second and third leg were transformed towards the identity of the first leg. Given that normally, in males, only the first leg pair depicts so-called sex combs, the fact that in the mutant all legs developed this morphological marker pointed to a deregulation of the underlying segmental identity imposed by the homeotic master regulators. Over time, a number of other genes with a similar phenotype were isolated and classified as "*Polycomb* group (PcG)" genes. Their phenotype indicated that they all act as repressors of homeotic master regulators. Interestingly, early patterns of homeotic gene expression were not disturbed but they were only at later stages (Struhl and Akam 1985). This suggested that the products of the PcG were not responsible for establishing the segment-specific expression patterns of master regulators, but for maintaining them appropriately expressed for the rest of development. Additionally, mutations in PcG genes also increased the rate of transdetermination events observed in transplantation experiments, suggesting that this group is part of the system maintaining cellular identities (Grossniklaus and Paro 2014).

In *Drosophila*, mutations like *Trithorax*, which mimic loss-of-function homeotic mutations, were identified. Further genetic analyses also revealed the existence of an entire group of genes acting as repressors of the additional sex combs phenotype found in PcG mutants. Mutations in these genes reversed or redirected the phenotypic transformations of segmental structures in an opposite manner compared to the homeotic transformations observed in PcG mutants. The genes belonging to this class were termed "*Trithorax* group (TrxG)" and their products seemed to act as antirepressors that ensure a faithful expression activity of the master regulators (Kingston and Tamkun 2014).

While originally identified as regulators of homeotic genes in *Drosophila,* both genes of the PcG as well TrxG were subsequently found to be evolutionary highly conserved in plants and animals. Primarily by molecular homology, members of the two groups were identified in the genomes of a number of model organisms. In some way, this is not surprising as cellular memory is an inherent requirement for a multicellular organism. However, subsequent analyses found a much broader regulatory regimen for PcG. They not only required to maintain developmental fates, but also used in a large number of other transcriptional control processes requiring maintenance of stable and heritable gene repression. The representatives of the TrxG are molecularly and functionally a rather diverse group and were found to be not only needed for many active transcriptional processes but also other nuclear processes requiring chromatin decompaction or activation. PcG and TrxG factors do not shape decisions for gene regulation, which is rather implemented by transcription factors (TFs) acting as master regulators. However, in many cases these TFs are switched off or take-up new tasks after their initiating signal. The further maintenance of the active or repressed state and the transmission of this information to the next cell generation is achieved by the combined action of PcG and TrxG. Hence, *"cellular memory"* could be understood as a mechanism for maintaining *"transcriptional memory"*. Maintaining a transcriptional state in the absence of the initial trigger is a classical epigenetic task fulfilled by these factors. Indeed, subsequent studies using the corresponding technical advantages of the different model organisms under analysis, revealed a class of gene products with major activities in shaping and modulating chromatin structure and function in the nucleus.

3.3 Biochemical Characterization and Molecular Function of PcG/TrxG Proteins

The molecular characterization of the genes and products of the PcG and TrxG provided deep insights into their specific role in the nucleus. General chromatin research had identified the important role of histones not only as packing material for the genome but as an additional carrier of information (see book ▶ Chap. 1 of Wutz). Histones depict an elaborate pattern of covalent modifications of their amino acids. Specific histone modifications were found to be correlated with defined transcriptional states, providing a platform for a gene-specific control and, therefore, producing an apparent code in addition to the genetic information encoded in the sequence of the DNA. An elaborate machinery consisting of chromatin factors writing, reading, and, if necessary, erasing the histone marks was revealed over the years (see book ▶ Chap. 1 of Wutz). Strikingly, among these so-called writers, readers, and erasers of epigenetic information, many were found to be encoded by PcG and TrxG genes.

Another interesting result was the discovery that PcG genes and to some degree also TrxG genes were genetically interacting. Subsequent molecular characterization revealed an interaction and cooperation of the proteins encoded by these groups in large and biochemically defined multimeric complexes. PcG proteins form two major types of complexes, *Polycomb* Repressive Complex 1 (PRC1) and 2 (PRC2) (Kuroda et al. 2020). Four PcG members constitute the core PRC1 complex in *Drosophila*: Polycomb (Pc), Polyhomeotic (Ph), Posterior sex combs (Psc), and Sex combs extra (Sce), also known as dRing1. PRC2 comprises four core proteins: Enhancer of zeste (E(z)), Suppressor of zeste 12 (Su(z)12), Extra sex combs (Esc), and Chromatin assembly factor 1, p55 subunit (Caf1-55), also known as p55 or Nurf55.

PRC1 recognizes and binds to the repressive H3K27me3 mark via the chromodomain of Polycomb and blocks chromatin remodeling and transcription. PRC1 has also E3 ubiquitin ligase activity mediated by Sce/dRING1, which monoubiquitylates H2AK119 (in Drosophila the corresponding K118 is ubiquitylated). Through E(z)'s methyltransferase activity, PRC2 trimethylates lysine 27 on histone H3 (H3K27me3) at genes targeted for silencing, thus being responsible for this distinctive histone mark of PcG system-mediated repression. The Su(z)12 and Esc subunits enhance this histone-modifying activity. The composition of the core complexes is conserved also in mammals (◘ Fig. 3.1). There, to deal with the higher regulatory complexity of the organism, a number of additional orthologs and factors cooperate with the core subunits to execute cell type- and tissue-specific functions. The concerted action of these protein complexes results in the stable and heritable transcriptional repression of PcG target genes.

The genetically defined group of TrxG proteins on the other hand was found to comprise a large and more divergent family of protein functions and complexes involved in the activation of transcription. They counteract PcG-mediated silencing by utilizing different mechanisms including the posttranslational modifications of histones, chromatin remodeling, recruitment of the transcriptional machinery, and

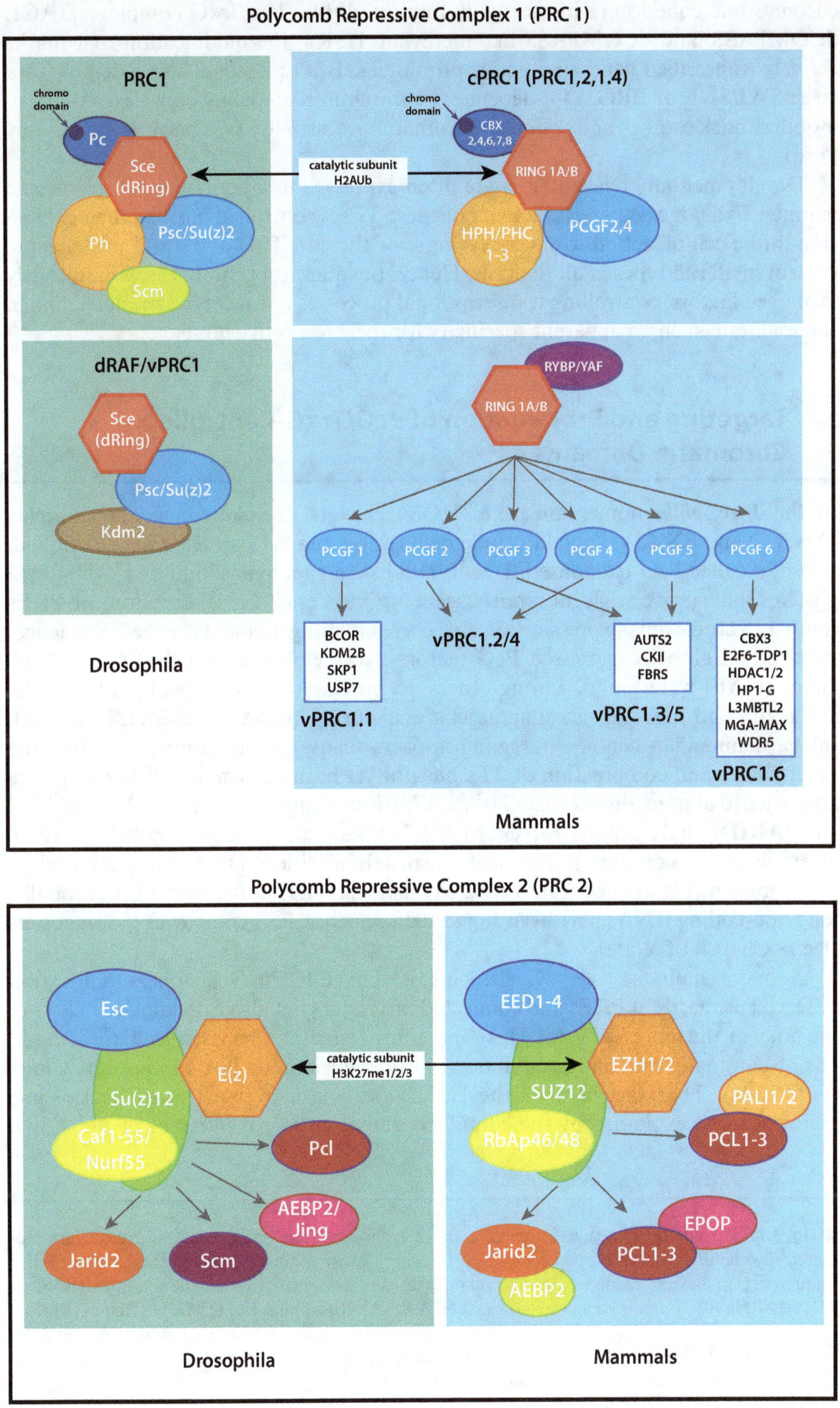
Polycomb Repressive Complex 1 (PRC 1)
PRC1
chromo domain
Pc
Sce (dRing)
Ph
Psc/Su(z)2
Scm
catalytic subunit H2AUb
cPRC1 (PRC1,2,1.4)
chromo domain
CBX 2,4,6,7,8
RING 1A/B
HPH/PHC 1-3
PCGF2,4
dRAF/vPRC1
Sce (dRing)
Psc/Su(z)2
Kdm2
Drosophila
RYBP/YAF
RING 1A/B
PCGF 1
PCGF 2
PCGF 3
PCGF 4
PCGF 5
PCGF 6
BCOR KDM2B SKP1 USP7
vPRC1.1
vPRC1.2/4
AUTS2 CKII FBRS
vPRC1.3/5
CBX3 E2F6-TDP1 HDAC1/2 HP1-G L3MBTL2 MGA-MAX WDR5
vPRC1.6
Mammals
Polycomb Repressive Complex 2 (PRC 2)
Esc
Su(z)12
E(z)
catalytic subunit H3K27me1/2/3
Caf1-55/ Nurf55
Pcl
AEBP2/ Jing
Jarid2
Scm
Drosophila
EED1-4
SUZ12
EZH1/2
PALI1/2
RbAp46/48
PCL1-3
EPOP
Jarid2
PCL1-3
AEBP2
Mammals

chromosome cohesion (Kingston and Tamkun 2014). The TrxG complexes TAC1, dCOMPASS, and dCOMPASS-like methylate H3K4, the modification that marks actively transcribed genes and active promoters. TrxG members also assemble into large SWI/SNF or BRG ATP-dependent chromatin remodeling complexes that can mobilize nucleosomes and promote chromatin accessibility (see book ▶ Chap. 2 by Paro).

Developmentally relevant TFs are often expressed in a tissue- and stage-specific manner. To a large degree, the activity of these TFs is controlled through their expression in the cell of action. Conversely, most of the core PcG and TrxG proteins are present in all cell types at all the time. Hence, the question is how these ubiquitously expressed factors, controlling transcriptional processes in a cell type specific manner, can maintain such an apparent specificity for their respective target genes.

3.4 Targeting and Propagation of PcG/TrxG-Controlled Chromatin Domains

In the *Drosophila* homeobox gene (HOX) clusters, the Antennapedia Complex (ANT-C) and BX-C, *cis*-regulatory elements attracting PcG proteins were identified. These so-called PcG Response Elements (PREs) can act in isolation by binding PcG proteins that silence adjacent reporter genes (Simon et al. 1993). Deletions of PREs result in ectopic mis-expression of the corresponding homeotic genes. Sequence-specific and globally expressed PcG factors like Pleiohomeotic (Pho) and GAGA factor (GAF) were found, among others, to provide the bridge between the DNA and PRC1 and PRC2, generating target specificity (Kuroda et al. 2020) (◘ Fig. 3.2). In the mammalian genome, targeting appears to use similar principles, albeit the exact nature and combination of TFs has not yet been determined. PcG complexes are enriched at unmethylated CpG islands. Various components, like KDM2, SUZ12 and JARID2, have affinity for GC-rich DNA sequences which may help to drive assembly of the complexes at nucleation sites. In addition, DNA methylation plays an important role in anchoring PRC2 to particular sites in mammals. Additionally, long non-coding RNAs have been suggested to anchor PcG complexes to target sites (see ▶ Chap. 4 of Wutz).

A major hallmark of PcG silencing is the H3K27me3 histone modification observed primarily at PREs and connected promoters but also extending largely over the body of the repressed gene. This “spreading” effect can be visualized using methods of chromatin immunoprecipitation (ChIP, see Methods Box 1 in book ▶ Chap. 1 of Wutz). The placement of the H3K27me3 mark over extended nucleosome regions can be explained by the combined and cooperative activities of individual

◘ Fig. 3.1 Polycomb Repressive Complex 1 and 2 in flies and mammals. PRC1 and PRC2 are biochemically identifiable multiprotein complexes. The core factors are color coded to show the respective counterparts in flies and mammals. The two complexes have a distinctive enzymatic activity; dRING or RING1A/B in PRC1 places a monoubiquitin at H2AK118 (Drosophila) and H2AK118 (mammals) and E(z) or EZH1/2 place a methyl moiety on H3K27. Some of the core proteins are found in variant complexes (vPRC1.1-6), especially in mammals. In combination with additional accessory factors these complexes fulfill specific tasks in a tissue-specific manner

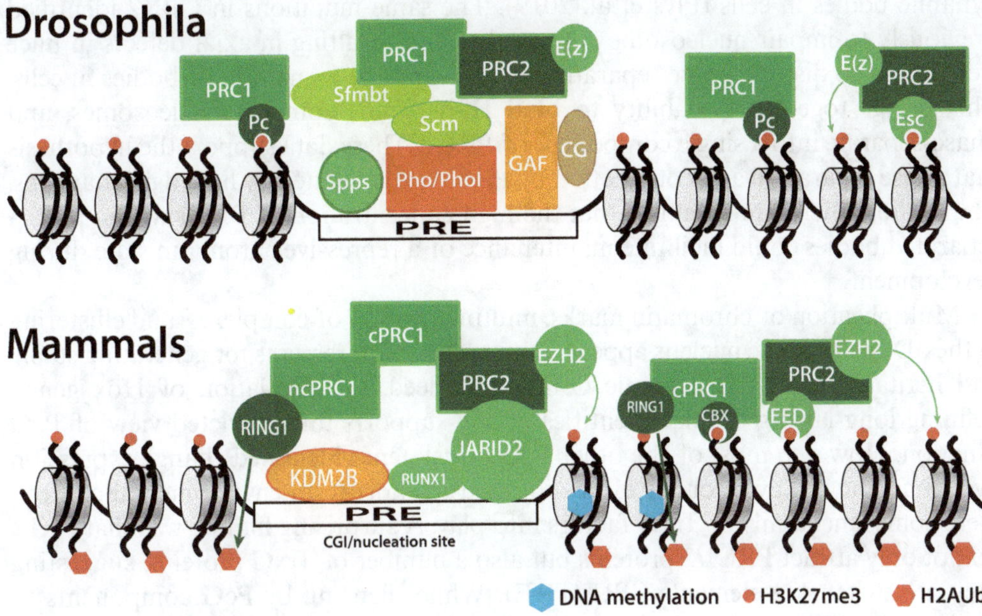

Fig. 3.2 Anchoring of PcG complexes to silencing elements. (top) In Drosophila PcG repressive elements (PREs) are recognized and bound by a combination of sequence-specific TFs (Spps, Pho/Phol, GAF and CG). They interact with PRC2 complexes which produce the H3K27me3 signal, attracting PRC1. ESC/EED of PRC2 and PC/CBX of cPRC1 bind H3K27me3 to drive self-propagation or spreading of the repressive mark. As a result, PREs can induce nucleosome compaction and inhibition of transcription by interaction with paused pol II. (bottom) In mammals the sequence requirement of PREs and corresponding TFs are less well understood. Unmethylated CpG islands attract PcG complexes through factors like KDM2, SUZ12 and JARID2 that have an affinity for CG-rich DNA. In contrast to Drosophila, where the role of H2Aub is not well understood, in mammals this mark facilitates recruitment of PRC2 via JARID2. PREs can also be used as platform for factors of the TrxG thereby acting as a switch for gene modulation

members of PRC1 and PRC2. The activity of the H3K27 histone methyltransferase E(z)/Ezh in PRC2 is stimulated by the other PRC2 component Esc/Eed that binds to H3K27me3, suggesting a feed-forward mechanism that could progress along the chromatin fiber. The Polycomb protein of PRC1 acts as a reader by binding the H3K27me3 mark through its *chromodomain*. In combination with other components of PRC1, like Polyhomeotic (Ph) containing multimerization modules (SAM domain), extended local clustering of chromatin sections are established. This leads to local nucleosomal condensation at regulatory sites and, in consequence, reduced DNA accessibility for TFs. Interestingly, such locally condensed regions can also interact with similar silenced regions in other parts of the genome, producing large, microscopically visible domains (Entrevan et al. 2016). An example for such long-range interactions can be found at the two *Drosophila* HOX clusters, normally separated by megabases on the linear chromosome, but colocalizing in the nucleus of cells where the Hox genes are repressed. The chemical property of certain components of the PcG induce the complexes to produce so-called phase-separated droplets, which have been termed PcG bodies. Such sub-nuclear compartments are not defined by molecular boundaries like membranes but, like oil and water, are separated by their disparate chemical properties. Recently, it was shown that the mammalian Polycomb component (CBX2) of canonical PRC1 can phase-separate *in vitro* and generate dynamic bodies in cells (Plys et al. 2019). The same mutations in CBX2 identified previously to impair nucleosome compaction and resulting in axial defects in mice were found to disrupt phase separation *in vitro* and the formation of bodies in cells. This brings together the ability to bind H3K27me3, compact nucleosomes, and phase-separate into a single component of PRC1. These data support the hypothesis that phase separation and compaction by PRC1 are generated by linked mechanisms. The authors therefore speculate that the high concentration of PRC1 within phase-separated bodies could facilitate maintenance of a repressive chromatin state during development.

Multiplication of chromatin marks, multimerization of complexes, and clustering in the 3D space of the nucleus appear to be central requirements for generating stably and heritably repressed genomic domains. Indeed, the regulation of Hox genes, defining long-lasting cellular identities, nicely supports this restricted view of PcG function. However, most of the other PcG target genes identified change expression over time, the cell cycle or are even subjected to short-term environmental control. Here comes the family of TrxG factors into play. A surprising finding was that PREs do not only attract PRC1/2 proteins but also a number of TrxG proteins, suggesting they are switchable elements (PRE/TRE). While silencing by PcG components to gene-specific PREs represents the default state, a competitive off/on switch and fine-tuning is imposed by particular combinations of TrxG proteins and transcription factors that react to developmental or environmental cues (◘ Fig. 3.3).

An interesting case in this respect are so-called bivalent domains. Originally discovered in mammalian embryonic stem cells, they recently were also identified in the *Drosophila* embryonic genome (Akmammedov et al. 2019). Bivalent domains are characterized by the concomitant presence, over an extended genomic region, of the characteristic PcG histone mark H3K27me3 and the typical TrxG mark H3K4me3 (see also Methods Box 1 in book ▸ Chap. 1 of Wutz). In early embryonic stages, this dual marking at genes appears to signal a transcriptionally poised condition. Hence, a gene "waits" for the appropriate developmental signals to drive its expression into an active or repressed state, depending on the fate of the cell lineage. With ongoing developmental time and programming, the signals are resolved into either a complete H3K27me3 pattern (and/or in the case of mammals with additional DNA methylation signals) when the gene is repressed or a complete H3K4me3 (and accompanying H3/H4 acetylation) when the gene is active (Voigt et al. 2013). This site-specific anti-repressor capability of members of the TrxG provides a means for how localized competition can switch the expression state from silenced to activated, also known as the "bivalent master switch" model (Kuroda et al. 2020). A central aspect is a shift of the balance by the addition of acetylation moieties, opening up nucleosomal condensation and allowing transcriptional activation. Indeed, PRC1 proteins have been found complexed with acetyltransferases and acetyl-binding bromo domain-containing proteins. Hence, the local balance between PcG and TrxG proteins defines the state of histone modifications and, consequently, the accessibility and processability of the chromatin template. However, based on theoretical considerations, it was suggested that poised chromatin is produced by a bistable state imposed by the PcG/TrxG competition rather than through a true bivalent marking of nucleosomes (Sneppen and Ringrose 2019). In their model, poised chromatin is bistable and not bivalent. They suggest that bivalent chromatin containing H3K4me3 and H3K27me3 is present as an unstable background population in all system states, and different subtypes co-occur

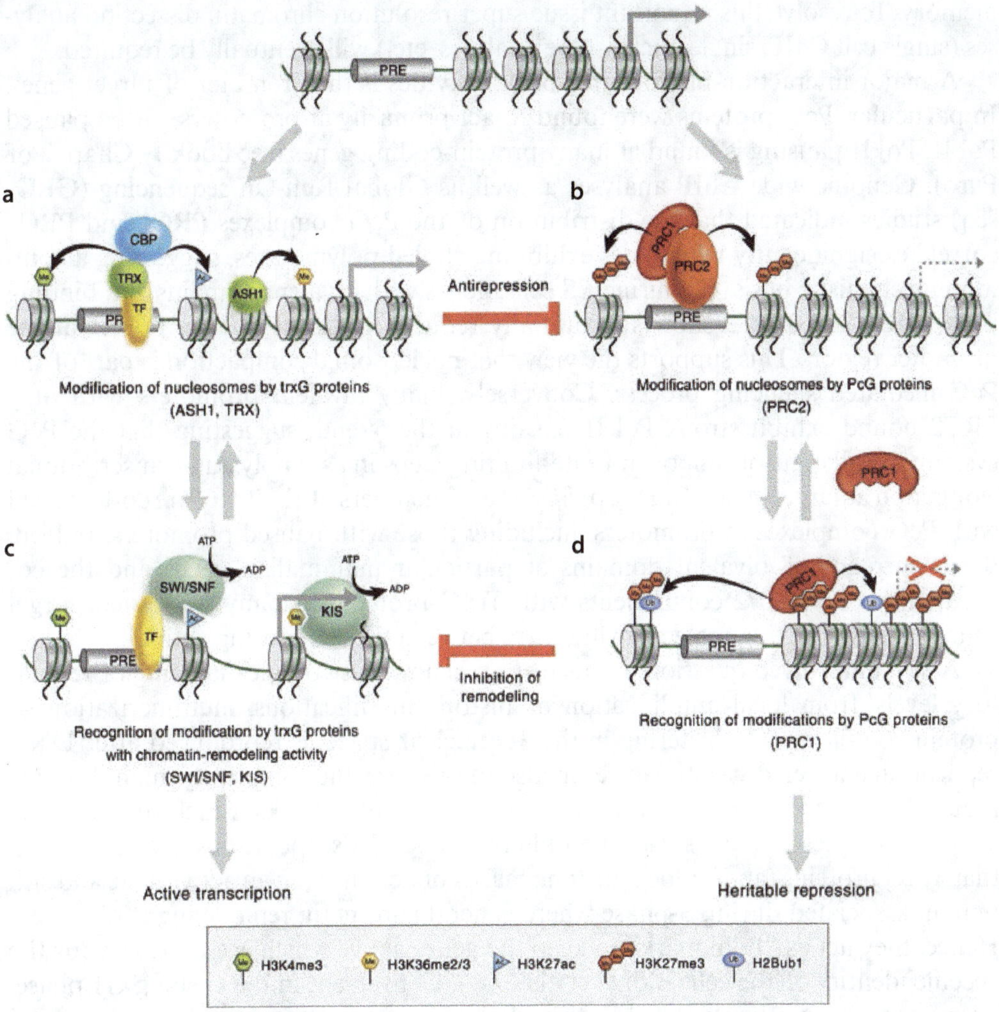

Fig. 3.3 Antagonistic functions between TrxG and PcG factors. Both TrxG and PcG families include proteins that covalently modify histones and those that non-covalently modify chromatin. Covalent modifications on histones can promote or block the binding or activity of TrxG complexes (e.g., SWI/SNF and KIS), PcG complexes (e.g., PRC1 and PRC2), or other factors involved in the maintenance of active or repressed states. Binding by these latter complexes has the potential to lead to further covalent modifications, thus leading to iterative cycles of covalent modification and recognition of the covalent marks. **a** TrxG factors are attracted to PREs at promoter regions and mark histones for activity. **b** In a different cell type the same gene is marked by negative histone modifications by PcG complexes bound at PREs. **c** Nucleosome remodeling complexes recognize the active histone marks and open nucleosomal packaging for access to the transcriptional machinery. **d** Modification of histones by PcG complexes result in compaction of nucleosomes and in gene repression. Depending on local constitution and concentration of the antagonistic factors, genes are either kept in an active or permanently repressed form. (from (Kingston and Tamkun (2014))

with active and silent chromatin. In contrast, bistability, in which the system switches frequently between stable active and silent states, occurs under a wide range of conditions at the transition between monostable active and silent system states. The question whether bistability and not bivalency is associated with poised chromatin has implications for the understanding of how the PcG/TrxG system maintains cellular memory. To resolve this important issue super resolution chromatin dissection analyses (single cell ChIP, single nucleosome analyses, etc.) will eventually be required.

A major interaction platform for both activities is the promoter of target genes. In particular, PcG proteins were found to act primarily at promoters with a paused Pol II. Pol II pausing is found at many protein-coding genes (see book ► Chap. 2 of Paro). Genome-wide ChIP analyses as well as Global Run-On sequencing (GRO-seq) studies indicated that the distribution of the PcG complexes PRC1 and PRC2 correlates significantly with genes exhibiting stalled polymerases, suggesting a common mechanistic basis. In murine ES cells, genes with bivalent domains and binding both PRC1 and PRC2 show substantially reduced levels of paused Poll II at the promoter region. This supports the view that nucleosomal compaction is part of the PcG-mediated silencing process. Conversely, many bivalent promoters with only PRC2 bound exhibit strong Pol II pausing at the 5′-end, suggesting that the PcG system overall controls important rate-limiting steps in assembly and transcriptional elongation at target genes. In *Drosophila*, many members of the TrxG are co-localized with PcG complexes at promoters, including those with paused promoters. Indeed, the occurrence of bivalent domains at particular mammalian genes and the co-occupancy of PRC1/2 components with TrxG proteins at many *Drosophila* target genes suggest a strong molecular interface between these opposing activities.

A still unresolved question in this respect is how this complex layering of regulatory levels, from local amplification of histone modifications, multimerization of protein complexes, to clustering in the 3D nuclear space, is reproduced after DNA replication and cell division. However, also in this case, the complexity might become reduced to the maintenance of a few relevant molecular marks, which subsequently act as entry sites to reconstruct the full epigenetic landscape. It has been suggested that TrxG proteins (and/or local histone marks or activity) tag an active gene and still remain associated during S-phase where general chromatin remodeling is observed. Hence, they act as "bookmark" to keep the genes active which are necessary for the specific identity of the cell (Blobel et al. 2009). Conversely, in the ensuing G1 phase, all genes without a bookmark become silenced again by default and by the tagging through a PRE(s). This would ensure that only one part of the complex transcriptional memory machinery is required to be faithfully duplicated and actively transported through DNA replication and mitosis.

3.5 Switching Memory and the Role of Non-coding RNAs

The example of bivalent chromatin domains showed how PcG/TrxG competitive interactions lead to permanent opposing expression states necessary for developmental stability. However, many repressed genes need to become activated or active genes repressed at later developmental times or when required by physiological or environmental conditions. An example for the latter is the process of vernalization in plants (Costa and Dean 2019). Since plants are sessile, they need to adjust their growth, development, and physiology to the changing environment. To ensure that reproductive development only starts after the winter period and, therefore, during more favorable environmental conditions, plants developed mechanisms to "remember" that they experienced prolonged periods of cold. Central to this process is the epigenetic silencing of a floral repressor during cold periods. In *Arabidopsis thaliana*, the

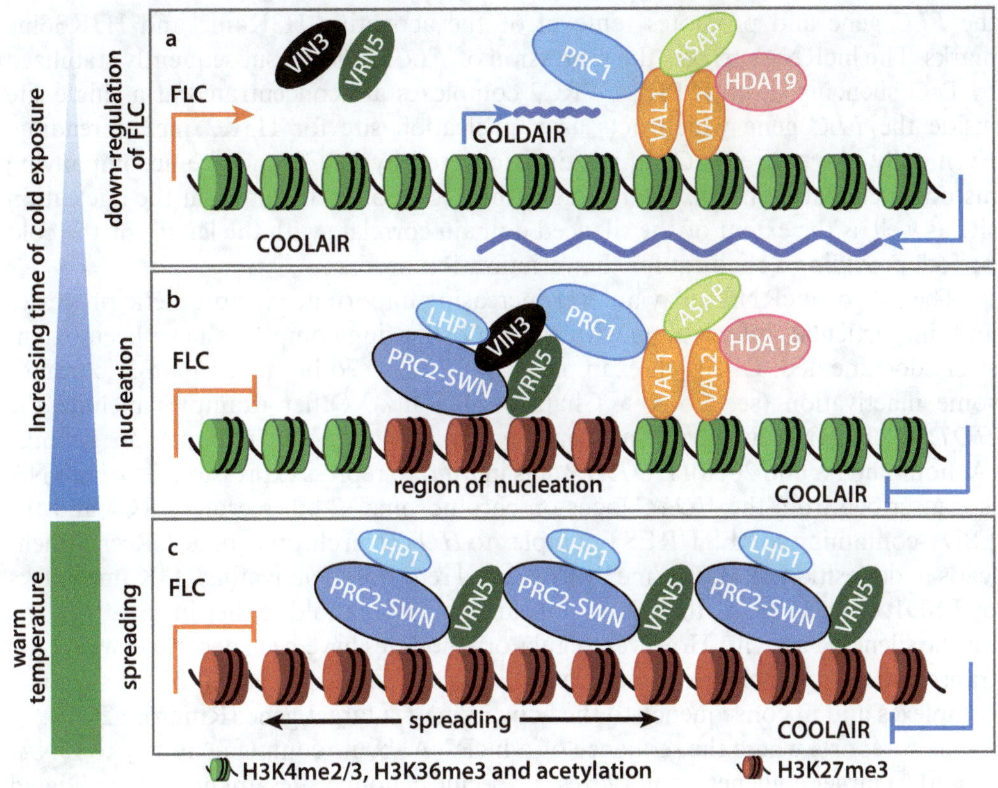

Fig. 3.4 Model for a lncRNA inducing an epigenetic switch at FLC. **a** Before cold exposure during winter time the FLC locus is expressed and the histones marked with H3K4me2/3, H3K36me3 and acetylation. At the start of cold exposure, the expression of the internal lncRNA Coldair attracts PcG complexes to the gene. Additionally, the anti-sense lncRNA Coolair becomes transiently expressed at high levels inducing the down-regulation of FLC transcription. Near the nucleation region VAL1/2 attract the histone deacetylase HDA19, the transcriptional regulator complex ASAP and PRC1, linking transcriptional repression to PcG silencing. **b** Nucleation start by ASAP associating with PRC2 accessory proteins VIN3 and VRN5 and with the PRC2-SWN complex. This generates a metastable silenced state of FLC starting accumulation of H3K27me3 signals at the nucleation region. **c** After exposure to warm temperatures, this signal together with PRC2-CLF and LHP1 spreads over the entire locus establishing a long-term epigenetically silenced state. This condition is maintained across multiple cell division until reset in the next generation during embryogenesis

FLOWERING LOCUS C (FLC) transcription factor acts as a break to flowering. The expression of this repressor is gradually reduced during long periods of cold, so that, once plants detect inductive photoperiods and warm temperatures, the switch to flowering can be activated. The repression of *FLC* is stable and maintained for many months after cold exposure until embryogenesis in the seed producing the next generation, leading to reactivation of *FLC* (Fig. 3.4).

The discovery of a pair of long non-coding RNAs (lncRNAs) transcribed at the *FLC* locus revealed a novel role for such transcripts in an epigenetic switch mechanism. Cold induces the local accumulation of both lncRNAs: *COLDAIR*, which is transcribed inside an intron of *FLC*, and *COOLAIR*, which is transcribed antisense to the *FLC* mRNA. *COLDAIR* recruits plant PRC2 containing the E(z) homologs CURLY LEAF (CLF) and SWINGER (SWN). This results in enrichment of H3K27me at the 5′ and 3′ ends of *FLC*. The antisense *COOLAIR* transcript spans the *FLC* gene and promotes removal of the activating H3K4me and H3K36me marks. The lncRNAs trigger the repression of *FLC*, which is subsequently stabilized by PcG silencing. Like at PREs, PRC2 complexes are concentrated at a single site inside the *FLC* gene which acts as a nucleation site for H3K27me3 spreading. Eventually, the entire *FLC* chromatin is covered by PRC2 and the accompanying histone methylation mark. Interestingly, the degree of methylation at the nucleation site as well as the extent of the silenced domain correlate with the length of the cold period, providing a rheostat for the vernalization process.

The role of lncRNAs play an ever-increasing importance in epigenetic processes and, in particular, interactions with the PcG silencing complexes have been extensively documented. One such example is XIST involved in mammalian X chromosome inactivation (see book ▸ Chap. 4 of Wutz). Other examples include the *HOTAIR* lncRNA involved in silencing *Hox* genes in the mammalian genome (Achour and Aguilo 2018). *HOTAIR* acts in *trans* to repress expression. The lncRNA is transcribed from the *HOXC* locus on chromosome 12 but recruits PRC2 and the LSD1-containing coREST/REST complex to *HOXD* on chromosome 2. Recruitment leads to deposition of H3K27me by PRC2 and removal of activating H3K4me marks by LSD1. In these cases, the RNAs appear to act as scaffolds either in *cis* or in *trans* for the silencing activity. However, regulatory modules have been described where the transcription of a lncRNA through a PRE leads to the displacement of the silencing complexes and in consequence to the activation of a target gene (Ringrose 2017). In animals, reports where the sequence of a lncRNA, by its complementarity to DNA, is used to target epigenetic complexes to specific genomic sites are not well validated and still controversial. This is in contrast to small RNAs in plants that can clearly recruit chromatin and DNA modifying complexes to target sites in a sequence-specific manner (see book ▸ Chap. 6 of Grossniklaus).

In summary, there are several cases where the act of transcribing a lncRNA (or any other transcript) can displace a PcG silencing complex. In parallel, transcripts exist which could locally attract PcG complexes in cis and induce the silencing of a gene. This offers the potential of a switch mechanism, based on the activity of RNAs. Indeed, the role of lncRNAs acting in *cis* in switching PcG/TrxG transcriptional states has been modelled based on published criteria (Ringrose 2017). Depending on the kinetic parameters, lncRNAs can act to recruit PRC2 to target chromatin or can be induced to evict PRC2 from silenced chromatin (see ◘ Fig. 3.5).

3.6 Losing Memory

A natural form of cellular memory loss is required during the transmission of genetic information by the germ cells for fertilization and generation of progeny. Germ cells need to strap any epigenetically encoded information covered by tissue-specific chromatin domains or epigenetically imprinted marks caused by environmental cues (Reik and Surani 2015). Indeed, development of germ cells encompasses several bottlenecks where chromatin information is apparently erased to ensure a faithful re-start of transcriptional programming at fertilization. These range from the replacement of histones with protamines, followed by a strong compaction of the genome in sperm cells, to the complete genome-wide DNA demethylation observed in early mammalian preimplantation embryos (see book ▸ Chap. 6 of Grossniklaus).

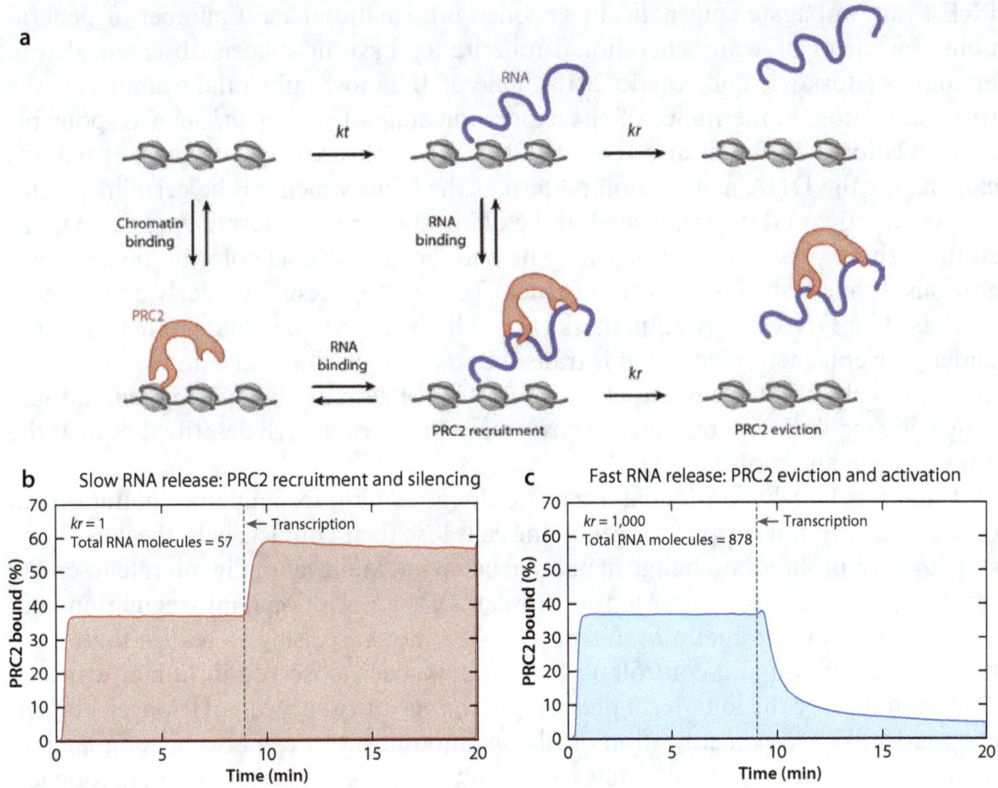

Fig. 3.5 Simple kinetic model for interaction of PRC2 with RNA and chromatin. **a** The model is based on ordinary differential equations and contains three components: chromatin (gray), PRC2 (red), and RNA (dark blue). (Left) Chromatin binding: PRC2 can bind directly to chromatin. The dissociation constant (Kd) for this interaction is 86 nM, measured for PRC2 binding to H3K27me3 modified histone tails. (Middle) RNA binding: PRC2 can also bind to RNA. The Kd used for this interaction in the simulations shown is 30 nM. RNA is produced with rate kt, released from chromatin with rate kr, and degraded with rate kd (not shown). (Right) RNA is released from chromatin after transcription, either unbound (top) or bound (bottom) by PRC2. **b**, **c** The system was simulated over a time course, and RNA transcription was initiated after 8 min. The input quantities for these simulations were set to reflect average molecule numbers in a 0.1 μm^3 volume in a *Drosophila* embryo nucleus at cycle 12, based on total nuclear concentrations (approximately the confocal volume used for fluorescence correlation spectroscopy experiments. The quantities were PRC2 = 5 molecules; H3K27me3 = 52 molecules, assuming 6% of tails modified; kt = 0.2 s^{-1}; and kd = 0.01 s^{-1}. kr was varied as shown in panels b and c. RNA molecules were assumed to be localized in the volume of the simulation. (From Ringrose (2017))

Interestingly, exceptions have been observed in both animal and plant reproduction where not all epigenetic information is erased, influencing the normal growth and physiology of the progeny. A first case was described in *Drosophila* with a PRE controlling a *white* reporter gene (responsible for the red eye color in flies) (Ciabrelli et al. 2017; Cavalli and Paro 1998). Normally, the PRE would silence the reporter resulting in white-eyed flies, given that the endogenous *white* locus was mutant. Through a single transcriptional switch, the reporter gene could become activated and stay on through the entire development of the fly. Surprisingly, the activated form was not erased in the germ line but transmitted to the progeny, which also displayed red eye color. The effect could be maintained over several generations, demonstrating that, under specific conditions, a single transcriptional signal, changing the modularity of

PREs, can propagate epigenetically encoded information over a number of generations. Instances of transgenerational inheritance have now been observed also in mammals (Boskovic and Rando 2018). One of the most influential examples is the transgenerational inheritance of the expression state of the *agouti* gene, responsible for coat color, in mice (see also Book ► Chap. 9 of Santoro). Specific diets of parents can change the DNA methylation pattern at the locus which, in the germline, is not always erased and thus propagated to the offspring. The transferred epigenetic mark controls the expression of the *agouti* gene and, hence, the coat color in the next generations. While transmission through meiosis, gametogenesis, and early embryogenesis based on DNA methylation marks are well documented, in many other cases the underlying epigenetic signal that is transmitted and causes an effect in the progeny is not always clear (Grossniklaus et al. 2013). Conversely, in plants, transgenerational epigenetic regulation in response to environmental stress is well described both at the physiological and molecular level.

Cancer is largely considered a genetic disease. With accumulating mutations in oncogenes or tumor suppressor genes, the cells lose their contacts with the neighbors, start to over-proliferate, change metabolic behavior, and eventually migrate to create the deadly metastases. Cancer is also a disease of transcriptional mis-regulation and, given the role of chromatin in this process, it is not surprising to realize that many mutations in chromatin-controlling factors have been observed in human cancers. Mutations ensure the long-term phenotypic change of cancer cells. However, epigenetic marks like DNA methylation or histone modifications can also be long-lasting, maintaining gene expression states over many cell generations. In consequence, deregulated expression in cancer cells need not only have a mutational basis but is often driven by epigenetic mis-regulation. In ► Chap. 9 cancer epigenetics is described in great detail. Here, two examples for mutations in components of the cellular memory system will be outlined, illustrating how the loss of cellular identity during ontogeny can lead to the development of a tumorigenic state.

The human TrxG gene Mixed Lineage Leukemia (MLL1) was initially identified as a chromosomal translocation found in blood cells and causing acute lymphoblastic or myeloid leukemia. MLL1 codes, as its *Drosophila* ortholog *trithorax* (*trx*), for a histone 3 lysine 4 methyltransferase (KMT), catalyzing the addition of the positive histone mark H3K4me3. The translocation brakes inside the gene, often fusing the N-terminal part of the MLL1 protein with other transcriptional control factors. The C-terminal part containing the KMT activity (SET domain) is removed, depriving MLL1 from its normal activating function. Depending on the fusion partner, MLL1 can repress or activate target genes in an aberrant manner and thereby induce cancer-typical expression profiles. Since the KMT activity normally repels the DNA methylation machinery, the MLL1 fusions, where this part is often absent, are often found at hypermethylated promoters. This results in the down-regulation of many target genes. Conversely, the N-terminal part of MLL1 contains a DNA-binding affinity. The recruitment of co-activators through the fused partner protein can result in the hyperactivation of target genes. The result is a highly deregulated gene expression profile found in leukemic cells. By studying the molecular partnerships of MLL1 fused, for example, to the AF9 activator, specific small molecule inhibitors could be identified. In one case, MLL1 interacts with the histone demethylase LSD1. When blocked by a small inhibitor (LSD1i) the aberrant activating function of the fusion MLL1 is reduced. Similarly, preventing another interaction partner, BRD4, to bind via its bromo domain to acetylated histones by blocking the reader function (BETi) also reduces the transcriptional hyperactivity (see also book ► Chap. 8 of Santoro). Hence, the detailed knowledge of the molecular processes encompassing TrxG-driven anti-repressor activity allowed the development of highly specific inhibitors, which are currently in clinical trials as anti-cancer drugs.

Mutations in PcG genes are often associated with a variety of human cancers of a large range of cell types and tissues. Similarly, down-regulation of members of the PcG protein complex in *Drosophila* can lead to neoplastic growth in imaginal discs (Beira et al. 2018). The role of the PRC1 component *polyhomeotic (ph)* has been studied in substantial detail, revealing how loss of cellular identity can drive cells to an over-proliferative behavior. Interestingly, no genetic instability is observed in these tumor cells. Epigenetic dysregulation reprograms the cells to an embryonic state and blocks them from entering differentiation. However, by enforcing development through the overexpression of a differentiation factor, tumorigenic competence is restrained (Torres et al. 2018). Not only in fly cancers but also in many human cancers, cells with overproliferative traits often depict a de-differentiated state. This reflects the characteristics and need of these cells to divide out of control, independent of the neighboring environment and developmental cues. Interestingly, novel approaches, termed "differentiation therapy", attempt to force cancer cells into the differentiation path again, eventually removing them from the pool of overproliferating cells. Indeed, a large number of new cancer drugs that are currently in clinical trials target epigenetic factors, attempting to reverse aberrant transcriptional profiles (see also book ► Chap. 8 of Santoro) (Audia and Campbell 2016).

In summary, the chromatin factors encoded by the PcG and TrxG are required for maintaining developmental and other transcriptional states. While decisions to acti-

Take-Home Message

- The PcG/TrxG system, including the chromatin modifying and remodeling activities, is evolutionary highly conserved and is involved in many basic cellular processes like maintenance of developmentally controlled gene expression states (cellular memory), mammalian X chromosome inactivation, and vernalization in plants.
- PcG multiprotein complexes are involved in long-term silencing of genes. H3K27me3 is a hallmark of PcG repression, correlated with condensed nucleosomal structures. PcG repressed domains cluster in the 3D space of the nucleus to produce stable silencing.
- TrxG complexes counteract PcG repression and are required for maintaining gene activity stable and heritable. H3K4me3 is the hallmark of the TrxG, which encodes writers and readers of histone modifications but also components of the nucleosome remodeling complexes involved in opening chromatin structures.
- Long non-coding RNAs play an important role in the silencing mechanism of the PcG. It has been suggested that lncRNAs interact with PcG protein complexes to add structural features, evict PRCs from PREs, or target the complexes to specific genomic sites.
- Deregulation of components of the PcG/TrxG system often results in cancer development. Epigenetic cancers are not dependent on the accumulation of genetic aberrations. Many new drugs against chromatin regulators are currently being tested in the clinic.

vate or silence a gene are made by master regulators, the resulting expression state can be maintained and faithfully inherited from one cell generation to the next by the epigenetic landscape imposed by either PcG- or TrxG-driven marks. The competitive and bistable nature of PcG and TrxG allows, however, the reversion of expression states if required, producing a highly versatile and adaptive regulatory network.

References

Achour C, Aguilo F (2018) Long non-coding RNA and Polycomb: an intricate partnership in cancer biology. Front Biosci 23:2106–2132

Akmammedov A, Geigges M, Paro R (2019) Bivalency in Drosophila embryos is associated with strong inducibility of Polycomb target genes. Fly 13(1–4):42–50. https://doi.org/10.1080/19336934.2019.1619438

Audia JE, Campbell RM (2016) Histone modifications and cancer. Cold Spring Harb Perspect Biol 8(4):a019521. https://doi.org/10.1101/cshperspect.a019521

Beira JV, Torres J, Paro R (2018) Signalling crosstalk during early tumorigenesis in the absence of Polycomb silencing. PLoS Genet 14(1):e1007187. https://doi.org/10.1371/journal.pgen.1007187

Blobel GA, Kadauke S, Wang E, Lau AW, Zuber J, Chou MM, Vakoc CR (2009) A reconfigured pattern of MLL occupancy within mitotic chromatin promotes rapid transcriptional reactivation following mitotic exit. Mol Cell 36(6):970–983. https://doi.org/10.1016/j.molcel.2009.12.001. S1097-2765(09)00904-6 [pii]

Boskovic A, Rando OJ (2018) Transgenerational epigenetic inheritance. Annu Rev Genet 52:21–41. https://doi.org/10.1146/annurev-genet-120417-031404

Cavalli G, Paro R (1998) The Drosophila Fab-7 chromosomal element conveys epigenetic inheritance during mitosis and meiosis. Cell 93(4):505–518

Ciabrelli F, Comoglio F, Fellous S, Bonev B, Ninova M, Szabo Q, Xuéreb A, Klopp C, Aravin A, Paro R, Bantignies F, Cavalli G (2017) Stable Polycomb-dependent transgenerational inheritance of chromatin states in Drosophila. Nat Genet 81:83. https://doi.org/10.1038/ng.3848

Costa S, Dean C (2019) Storing memories: the distinct phases of Polycomb-mediated silencing of Arabidopsis FLC. Biochem Soc Trans 47(4):1187–1196. https://doi.org/10.1042/BST20190255

Entrevan M, Schuettengruber B, Cavalli G (2016) Regulation of genome architecture and function by polycomb proteins. Trends Cell Biol 26(7):511–525. https://doi.org/10.1016/j.tcb.2016.04.009

Grossniklaus U, Paro R (2014) Transcriptional silencing by polycomb-group proteins. Cold Spring Harb Perspect Biol 6(11):a019331. https://doi.org/10.1101/cshperspect.a019331

Grossniklaus U, Kelly B, Ferguson-Smith AC, Pembrey M, Lindquist S (2013) Transgenerational epigenetic inheritance: how important is it? Nat Rev Genet 14(3):228–235. https://doi.org/10.1038/nrg3435

Hadorn E (1968) Transdetermination in cells. Sci Am 219(5):110–114. passim

Kingston RE, Tamkun JW (2014) Transcriptional regulation by trithorax-group proteins. Cold Spring Harb Perspect Biol 6(10):a019349. https://doi.org/10.1101/cshperspect.a019349

Kuroda MI, Kang H, De S, Kassis JA (2020) Dynamic competition of polycomb and trithorax in transcriptional programming. Annu Rev Biochem 89(1):235–253. https://doi.org/10.1146/annurev-biochem-120219-103641

Lewis EB (1978) A gene complex controlling segmentation in Drosophila. Nature 276:565–570

Plys AJ, Davis CP, Kim J, Rizki G, Keenen MM, Marr SK, Kingston RE (2019) Phase separation of Polycomb-repressive complex 1 is governed by a charged disordered region of CBX2. Genes Dev 33(13–14):799–813. https://www.genesdev.org/cgi/doi/10.1101/gad.326488.119

Reik W, Surani MA (2015) Germline and pluripotent stem cells. Cold Spring Harb Perspect Biol 7(11). https://doi.org/10.1101/cshperspect.a019422

Ringrose L (2017) Noncoding RNAs in polycomb and trithorax regulation: a quantitative perspective. Annu Rev Genet 51(1):385–411. https://doi.org/10.1146/annurev-genet-120116-023402

Simon J, Chiang A, Bender W, Shimell MJ, O'Connor M (1993) Elements of the Drosophila bithorax complex that mediate repression by Polycomb group products. Dev Biol 158(1):131–144

Sneppen K, Ringrose L (2019) Theoretical analysis of Polycomb-Trithorax systems predicts that poised chromatin is bistable and not bivalent. Nat Commun 10(1):2133. https://doi.org/10.1038/s41467-019-10130-2

Struhl G, Akam M (1985) Altered distributions of Ultrabithorax transcripts in extra sex combs mutant embryos of Drosophila. EMBO J 4(12):3259–3264

Torres J, Monti R, Moore AL, Seimiya M, Jiang Y, Beerenwinkel N, Beisel C, Beira JV, Paro R (2018) A switch in transcription and cell fate governs the onset of an epigenetically-deregulated tumor in Drosophila. elife 7:777. https://doi.org/10.7554/eLife.32697

Voigt P, Tee W-W, Reinberg D (2013) A double take on bivalent promoters. Genes Dev 27(12):1318–1338. https://doi.org/10.1101/gad.219626.113

Wolpert L, Tickle C, Martinez AC (2015) Principles of development. Oxford University Press, Oxford

4

Chromosome-Wide Dosage Compensation Systems

Contents

What You Will Learn in this Chapter

This chapter provides an introduction to chromosome-wide dosage compensation systems. We will examine the evolution of dosage compensation, which is thought to be driven by the appearance of differentiated sex chromosomes. In a subset of species with X chromosomal sex determination or XY sex chromosome systems, expression of X-linked genes is regulated by chromosome-wide modifications that equalize gene expression differences between males and females. The molecular mechanisms of X chromosome-wide dosage compensation have been studied in flies, worms, and mammals. Each of these species uses a distinct dosage compensation strategy with a different molecular mechanism. In the worm *Caenorhabditis elegans*, gene expression on the two X chromosomes of hermaphrodites is reduced to a level that approximates a single X chromosome in males. The fruit fly *Drosophila melanogaster* achieves dosage compensation by increased transcription of the single X chromosome in males to a level that is similar to the two X chromosomes in females. Lastly, in mammals, one of the two X chromosomes in female cells is transcriptionally inactive and a single X chromosome is transcribed in both sexes. Studies of dosage compensation systems provide insights into how epigenetic regulation controls gene expression and chromatin organization differentially within a cell.

4.1 Introduction: Evolution of Chromosome-Wide Dosage Compensation

Sexual reproduction can be observed in nearly all animal phyla. Different sex determination strategies have evolved, ranging from determination by environmental conditions including temperature to genetic sex determination with specific sex determination genes (Bachtrog et al. 2014). Sex determination systems allow the development of dimorphic phenotypes of males and females in a species. For this chapter, chromosomal sex determination is of primary interest as structural differences of sex chromosomes provide the context for the evolution of dosage compensation systems. In birds, males carry two Z chromosomes and females possess heteromorphic Z and W chromosomes. In mammals, males have an XY and females an XX sex chromosome constitution (Graves 2015). A wide range of XY, XO, and ZW sex chromosome systems have been documented in reptiles, amphibians, fish, and insects (Kaiser and Bachtrog 2010). The gene content of the sex chromosomes is different, though. For example, the snake and bird ZW systems are different and, hence, have evolved independently. Sex chromosomes in one species can correspond to autosomes in a different species, suggesting that sex chromosomes originate through modification of autosome pairs. How autosomes are selected to become sex chromosomes is just beginning to be understood. Evolution acts on sex chromosomes in a variety of ways that are species-, chromosome-, and context-dependent (Bellott et al. 2010). The general idea is that a sex determining locus arises on an ancestral autosome forming a proto-sex chromosome. Subsequent loss of sequences (erosion) on the proto-sex chromosome converts the ancestral autosome pair into a dimorphic sex chromosome pair (◘ Fig. 4.1).

The erosion of sex chromosomes has been explained as a consequence of recombination suppression. This could likely be facilitated by large inversions in the vicinity of the sex-determining region. Suppressed recombination facilitates the genetic

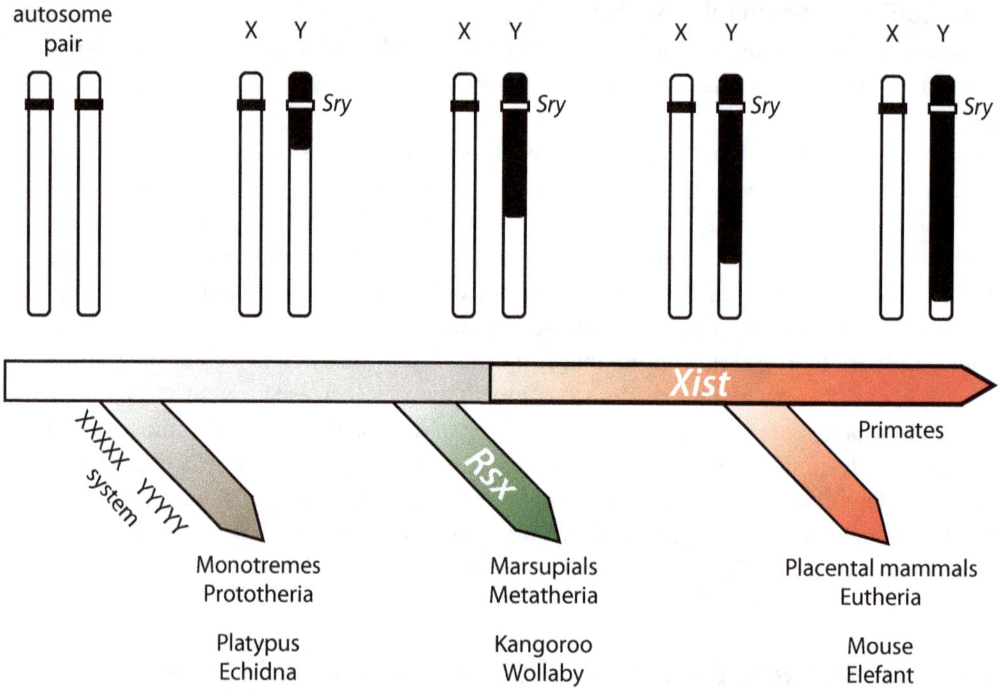

Fig. 4.1 Evolution of sex chromosomes in mammals. Placental and marsupial mammals possess XY sex chromosomes that have arisen from an ancestral autosome pair about 210 mya. These sex chromosomes correspond to autosomes in monotremes that possess a system of 5 X and 5 Y chromosomes. Evolution has shaped sex chromosomes by erosion and the addition of autosomal fragments to the sex chromosomes, leading to divergent chromosome structures between marsupials and placental mammals but also within placental mammals. Sequences lost due to erosion are marked in black along the evolving Y chromosome. As a consequence of the erosion of the Y chromosome, dosage compensation systems have arisen. Although placental and marsupial mammals inactivate the second X chromosome in female cells, the molecular details of the underlying mechanisms differ. In placental mammals, the *Xist* gene initiates X chromosome inactivation. In marsupials, the *Rsx* gene has emerged as the trigger by convergent evolution

linkage between the sex-determining gene and genes that are beneficial for one sex but detrimental to the other. For example, genes that are required for testis development reside on the mammalian Y chromosome together with the *sex determining region Y* (*SRY*) gene, which specifies male sex. Since females do not inherit a Y chromosome, there are no detrimental consequences from Y-linked genes to them. The XY sex chromosome system of placental mammals including humans is estimated to have evolved about 210 million years ago (mya) (Marshall Graves 2008). The mouse Y chromosome has lost all but a few genes that act in determining male sex and testis development whereas in other mammals, repeated addition of autosomal fragments to the XY chromosome pair has generated different and larger Y chromosomes. The evolution of different clades of mammals has led to divergent sex chromosome structures. Placental and marsupial mammals possess XY chromosomes that originate from the same ancestral autosome pair. The more distantly related monotremes are examples of species with multiple sex chromosomes (Fig. 4.1). Platypus and several species of echidna possess 5 X and 5 Y chromosomes, which are not homologous to the XY chromosomes in placental and marsupial mammals but show limited homology to the Z chromosome of birds (Graves 2015; Veyrunes et al. 2008). The human X chromosome is homologous to an autosomal region in monotremes. Further information on the evolution of mammalian sex chromosome systems can be found in a specialized review (Marshall Graves 2008).

4.1.1 Consequences of Gene Dosage Differences Arising from Sex Chromosome Erosion

Loss of genes on the sex-limited chromosome (the Y in mammals) and the evolution of dimorphic sex chromosomes can have two consequences. Firstly, a divergence from the ancestral gene dosage through erosion of the proto-sex chromosome relative to the ancestral autosomes is incurred. Whether an associated reduction of fitness[1] is tolerated or can be compensated for, remains to be resolved for each particular system. Secondly, a gene dosage difference between the two sexes arises as one of the sex chromosomes progressively shrinks in size and its gene content diminishes. It is likely that the process of erosion is constrained by deleterious effects caused by these gene dosage changes. Therefore, not all autosomes might be suitable for forming small sex chromosomes. In some instances, sex chromosomes appear to avoid erosion (Bachtrog et al. 2014), which might reflect a requirement to maintain ancestral gene dosage and prevents dosage compensation systems from arising.

Chromosome-wide dosage compensation systems have been discovered in worms, flies, and mammals (◘ Fig. 4.2). In some species, distinct processes are involved in adjusting dosage difference between the sexes and dosage differences between sex chromosomes and the ancestral autosome pair. This is the case in mammals, which possess a single active X chromosome in both sexes. Transcriptional silencing of one of the two X chromosomes in female mammals by the process of X chromosome inactivation approximates gene expression of the single male X chromosome. In this case, a reduction in fitness due to emerging differences in gene dosage between the sexes appears to have driven the evolution of dosage compensation. The mechanism of X to autosome (X:A) compensation remains to be resolved in mammals. In contrast, in the fruit fly *D. melanogaster* the regulation of the sex chromosomes is accomplished in a manner that balances gene dosage between the sexes and restores the level of X chromosome expression in males relative to the ancestral autosome pair. This is facilitated through the transcriptional hyperactivation of genes on the single X chromosome in males, such that an expression level is achieved that corresponds to two X chromosomes in females. Thus, the evolution of the dosage compensation system in flies appears to be driven by the adjustment of X-linked expression, such that it maintains gene expression levels between the sexes as well as the expression level of the ancestral autosome pair. In the worm *C. elegans*, the two X chromosomes

1 Reduction of fitness caused by shifts in sex chromosomal to autosomal gene dosage is consistent with studies on human monosomies, which show that inheritance of a single autosome in an otherwise diploid genome has high fitness costs. Human monosomies are generally detrimental and only monosomy X, and monosomies of smaller parts of chromosomes 5 and 1 are viable. Trisomies are also detrimental and only trisomy 13, 18, and 21 are viable in humans leading to Patau, Edwards, and Down syndrome, respectively.

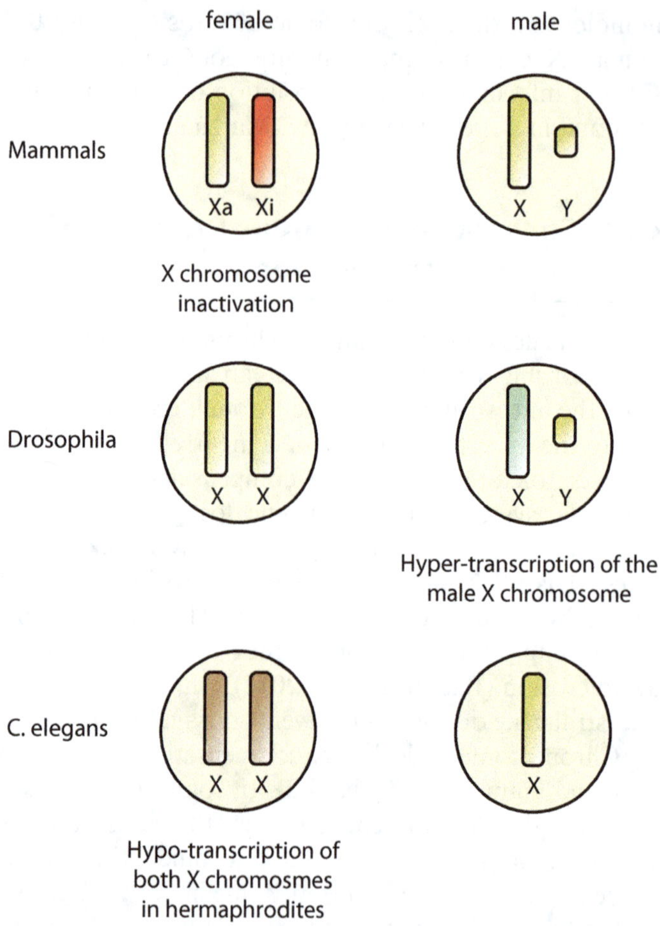

Fig. 4.2 Overview of dosage compensation in mammals, flies, and worms. X-linked gene expression differences between the sexes are compensated by different strategies. Mammals inactivate one of the two X chromosomes in female cells. In *D. melanogaster* the single male X chromosome is hyperactivated, and in *C. elegans* both X chromosomes are partially repressed in hermaphrodites

in hermaphrodites are partially repressed to approximate the gene expression level of the single X chromosome in males, which are XO. In *C. elegans*, it is again the dosage difference between the sexes that appears to drive the evolution of the dosage compensation mechanism. All studied dosage compensation systems to date have arisen within the context of an XY or XO sex chromosome system where the males carry the heteromorphic sex chromosome pair. Thus far, no chromosome-wide dosage compensation systems have been identified in ZW systems with heteromorphic female sex chromosomes. Consistent with the idea that X chromosomal dosage compensation in flies, worms, and mammals have evolved independently, these processes adopt molecularly distinct strategies and possess distinct genetic requirements.

4.2 The Dosage Compensation Complex of the Fruit Fly *Drosophila melanogaster*

The dosage compensation system of the fruit fly *D. melanogaster* has been extensively studied by genetic and biochemical approaches. *D. melanogaster* possess a XY sex chromosome system where a single X chromosome is present in male flies.[2] The single male X chromosome is the target of a dosage compensation complex (DCC) which acts to enhance transcription and, thereby, approximates the expression from two X chromosomes in female flies. In this manner, not only the difference in X-linked gene expression between the sexes but also the ancestral X:A ratio is restored. Thus, the single male X chromosome emulates an ancestral pair of two X chromosomes. Conceptually, the single X chromosome needs to be identified to target transcriptional upregulation. In addition, dosage compensation needs to be regulated to prevent aberrant hyper-transcription of the two X chromosomes in females. In *D. melanogaster*, dosage compensation and sex determination are regulated by a common upstream pathway that senses the number of X chromosomes. If a single X chromosome is present, this pathway diverges to regulate male sex differentiation and independently induces to the formation of the DCC.

The components of the DCC have been identified in genetic screens for mutations that prevent the development of male flies. The Male specific lethal (Msl) factors Msl1, Msl2, and Msl2 as well as Male lethal (Mle) and Males absent on the first (Mof) are proteins that contribute to the formation of the DCC. The respective genes are genetically required for male but not female development. Notably, the Msl2 protein is only present in males but not in females. Msl2 is regulated by the sex determination cascade and is essential for DCC assembly. Its absence in females prevents erroneous dosage compensation of both X chromosomes. Thus, Msl2 presents a molecular link between sensing the presence of a single X chromosome and assembly of the DCC. Once the DCC is assembled in male embryos, it is targeted to multiple sites on the single X chromosome. These sites have been referred to as chromosome entry sites as they appear to confer independent affinity for the binding of the DCC. These sites have further been associated with the DNA-binding protein CLAMP. CLAMP appears to be required for association of the DCC with chromosome entry sites and has also other functions in addition to its role in dosage compensation. Because mutations affecting CLAMP are detrimental to both male and female development, it could not be identified as a component of the DCC in the original genetic screens.

The DCC spreads out from chromosome entry sites over the entire X chromosome. DCC spreading is dependent on non-coding RNAs and the RNA helicase Mle. The discovery that functional RNAs are essential components of the *Drosophila* dosage compensation system was made serendipitously. Two genes encoding *RNA on the X 1* (*roX1*) and *rox2* were isolated as transcripts that are specifically expressed in males but not females. Their discovery was made considerably later than the identification of the protein components of the dosage compensation system. A likely reason why *roX1* and *roX2* were not discovered in genetic screens is that both RNAs are functionally redundant. Male development requires either *roX1* or *roX2*. Only combined mutations of *roX1* and *roX2* lead to a failure of dosage compensation and male lethality. It is thought that *Mle*, which encodes an RNA helicase, mediates the association of *roX1* and *roX2* with the Msl complex. Association of Mle and *roX*

2 Drosophila males do possess a Y chromosome that contains a small number of genes but it does not specify male sex. Therefore, the Y chromosome is not relevant for sex determination in flies.

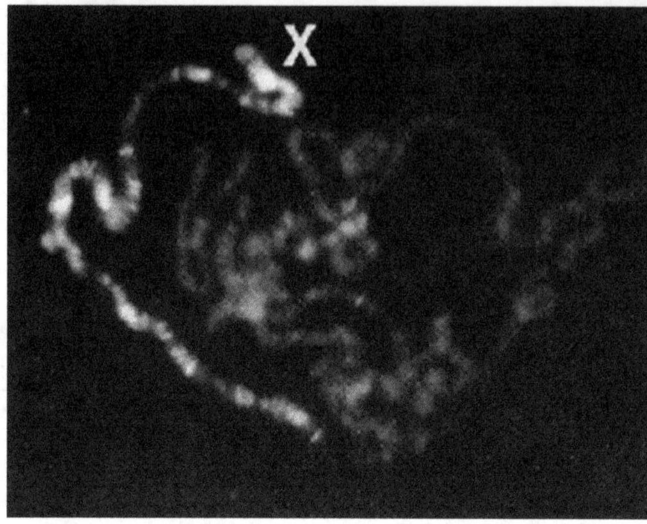

Fig. 4.3 Enrichment of H4K16ac on the X chromosome in *D. melanogaster* males. Immunostaining of a polytene chromosome spread with an antiserum specific for H4K16ac identifies the X chromosome in males. Polytene chromosomes are obtained from salivary glands and contain many strands of DNA due to endoreduplication of the genome in this polyploid tissue. This makes them well-suited for microscopy studies. (From Bone et al. (1994))

RNA thereby facilitates the spreading of dosage compensation from chromosome entry sites over the entire X chromosome.

Recruitment of the DCC to the X chromosome induces changes to chromatin and facilitates elevated transcription. An important component of the DCC that has been implicated in transcriptional regulation is the histone acetyltransferase Mof. Mof catalyzes the acetylation of histone H4 lysine 16 on the X chromosome in males (Fig. 4.3). The activity of Mof thereby contributes to the transcriptional activation of the single male X chromosome, consistent with the idea that acetylation of lysines in histone tails has an activating role for gene expression and facilitates accessibility of the DNA (see book ► Chap. 1 of Wutz, and ► Chap. 2 of Paro). The relevance of the observation that lysine 16 is specifically targeted by Mof for dosage compensation remains to be further explored. Furthermore, it appears that histone acetylation on the X chromosome is not stable or heritable by itself but depends on the continued recruitment of the DCC (see ► Box 4.1). Depletion of Mof or Msl2 by RNA interference in cultured male *D. melanogaster* S2 cells was shown to cause a loss of histone H4 lysine 16 hyper-acetylation on the X chromosome (Zhang et al. 2010). Although, the mechanism of dosage compensation in flies has been well characterized, our understanding is far from complete and, certainly, additional components of the dosage compensation pathway will be identified in the future. In this context, it is interesting to note that the kinase JIL1 was found associated with the DCC but its role in dosage compensation remains unclear.

Box 4.1: A Conceptual Experiment: Testing the Heritability of Chromatin Marks on the *D. melanogaster* Male X Chromosome (Box Fig. 4.1)

Depletion of MSL2 or MOF by RNA interference in S2 cells leads to loss of H4K16ac on X-linked gene promoters. **a** A Western analysis shows MSL2 and MOF protein abundance following RNAi-mediated depletion (*msl2* and *mof* RNAi) and control (Mock) transfected S2 cells. **b** A schematic representation of H4K16ac enrichment over genes from the X chromosome and the right arm of chromosome 3 as an autosomal control. H4K16ac enrichment was determined by ChIPseq (see book ▶ Chap. 1 of Wutz). The data are represented relative to DNA in the chromatin preparation to normalize for different sequence counts in different genomic regions. Each line represents a gene, the columns correspond to S2 cells as indicated on top of panel a. Genes enriched (yellow) and depleted (blue) for H4K16ac are indicated.

S2 cells were derived from male fly embryos and possess a single X chromosome that is subject to dosage compensation. Conversely, the X chromosome is

Box Fig. 4.1 Depletion of msl2 and mof in S2 cells

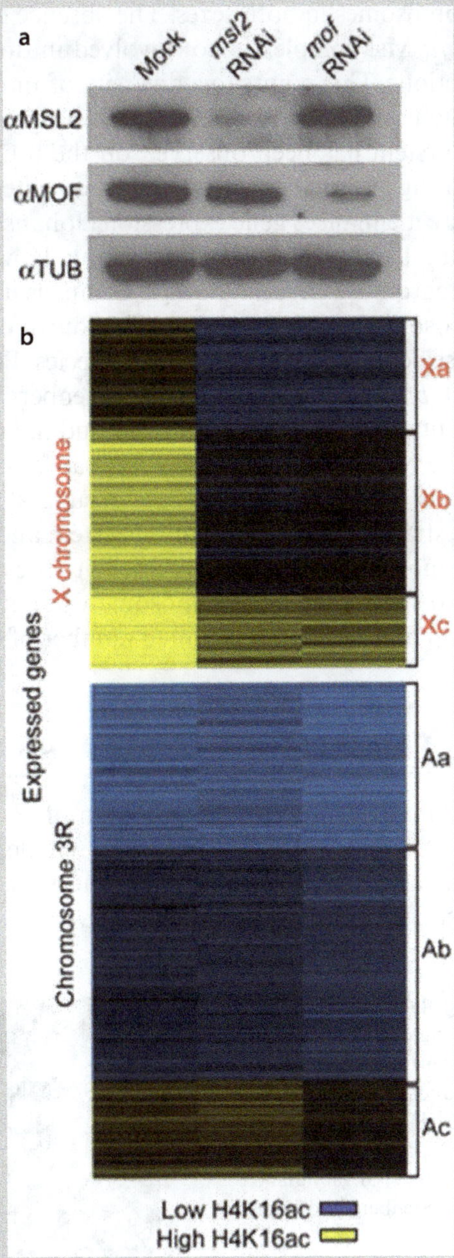

highly enriched for H4K16ac in control S2 cells. After depletion of MSL2 or MOF, this enrichment is lost and the X chromosome appears similar to chromosome 3R. This shows that the continuous activity of the DCC is required to maintain dosage compensation in S2 cells. (From Zhang et al. (2010))

The components of the *D. melanogaster* DCC have been investigated in distantly related insects to study the evolution of dosage compensation. An Msl complex is present in the distantly related insect *Sciara* (Ruiz et al. 2000) but localizes to all chromosomes in both sexes. The absence of sex-specific differences suggests that the *Sciara* Msl complex is not involved in dosage compensation and might have other functions. This points to a diversity of molecular mechanisms for dosage compensation among different insects. Interestingly, an independent chromosome-wide marking system has been observed on the fourth chromosome in *D. melanogaster*. The Painting of fourth (POF) protein associates with the fourth chromosome (◘ Fig. 4.4) where it enhances gene expression (Johansson et al. 2007). Initially, POF appeared to be unrelated to dosage compensation. However, a link between POF and an ancestral sex chromosome and dosage compensation system has recently been uncovered (Vicoso and Bachtrog 2013). The fourth chromosome of *D. melanogaster* is part of the sex chromosome in other fly species. POF is targeted to the male X chromosome in *D. busckii* and *D. ananassae* (Stenberg and Larsson 2011). In the latter, POF is recruited together with H4K16ac and the *Msl* complex. Overlap of chromatin marks of the X and fourth chromosome can be observed to varying degrees in different fly species. These observations could suggest a common origin of both pathways in an ancestral dosage compensation system and a potential separation of the X and forth chromosome in the lineage that led to the evolution of *D. melanogaster*. It is interesting to speculate that chromosome-wide modifications might be more common than would appear from the limited number of examples that are known today.

4.3 X Chromosome Inactivation in Mammals

In mammals, dosage compensation of X-linked genes is achieved by repression of one of the two X chromosomes in female cells. X chromosome inactivation ensures that, in both sexes, a single X chromosome remains active. This mode of dosage compensation might appear less intuitive than the *D. melanogaster* system as the ances-

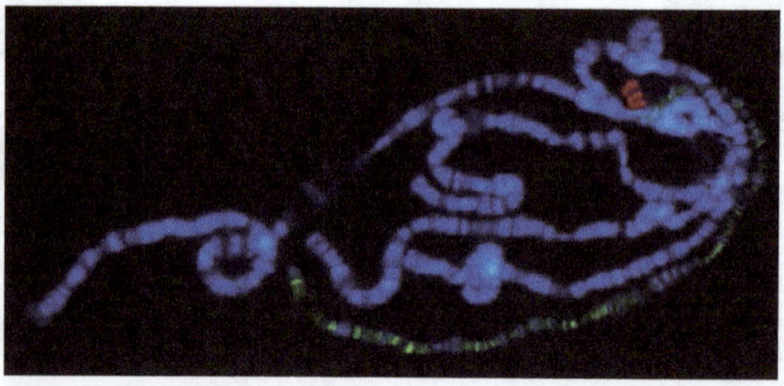

◘ **Fig. 4.4** Chromosome-wide marking of *D. melanogaster* X and fourth chromosomes. Immunostaining of POF (red) and Msl3 (green) of polytene chromosomes from *D. melanogaster* male salivary glands. (From Stenberg and Larsson (2011))

tral X:A gene expression ratio is changed from 2X:2A to 1X:2A. The relative X:A dosage shift associated with the evolution of the mammalian XY chromosome system could reduce fitness. In addition, mutations on the single X chromosome that is transcribed in both sexes are functionally hemizygous and cause phenotypes at a higher rate than recessive mutations on autosomes where complementation by the homolog is possible. Notably, random choice of which X chromosome is inactivated in placental mammals protects females from X-linked hemizygosity to some extent. Females comprise of a genetic mosaic of cells expressing either X-chromosome. Hence, in case of a heterozygous mutation, half of the cells will express the intact gene. However, it remains unknown to what extent the fitness costs arising from the evolution of the mammalian sex chromosome and dosage compensation systems have been absorbed. Although the evolution of X chromosome inactivation remains largely unclear, one potential scenario might be that it was derived from an extension of meiotic inactivation of unpaired chromosomes. Meiotic sex chromatin inactivation (MSCI) is a process that results in the transcriptional inactivation of unpaired chromosomes in meiosis during germ cell development. This process has been observed in a wide range of animal species. Inactivation is thought to allow unpaired chromosomes to be passed through meiosis. Reactivation of MSCI normally occurs after fertilization when embryo development starts. One might speculate that a perpetuation of MSCI of the paternal X chromosome might have evolved in mammals. As the paternal X chromosome is largely unpaired in male (XY) meiosis, silencing would be initiated by MSCI. In marsupial mammals, transcriptional inactivation of the paternally inherited X chromosome is observed, suggesting that imprinted X chromosome inactivation in marsupials might be based on mechanisms that extend silencing of MSCI. The process of random X chromosome inactivation in placental mammals poses a greater challenge to explain. Each female cell nucleus possesses two potentially genetically identical X chromosomes, whereby one is inactivated and the other remains active. Hence, both chromosomes assume divergent fates in the same nucleus. Therefore, a novel and complex mechanism had to be established at the split of placental and marsupial mammals.

4.3.1 The Mammalian Dosage Compensation Mechanism

The inactive X chromosome (Xi) was discovered serendipitously when a dark stained dot was observed in the nucleus of neurons from female cat brains (◘ Fig. 4.5). This dot was not present in male cat neurons (◘ Fig. 4.5). From this study, the Xi acquired its colloquial name “Barr body” in recognition of its discoverer (Barr and Bertram 1949). Subsequent studies revealed the identity of the Barr body as the Xi and the process of random X inactivation. A key insight into the mechanism was the identification of a transcript that is expressed only in female cells and encoded on the X chromosome. The *X inactive specific transcript* (*XIST*) gene was initially identified in humans and shown to produce a functional RNA that associates with the Xi. *XIST* is located within a region of the X chromosome that contains additional regulators of dosage compensation which together make up the X inactivation center (*XIC*). The *XIC* encompasses elements that signal the presence of an X chromosome and regulate *XIST* expression if more than one is present. *XIST* expression precedes the transcriptional repression of genes on the Xi. Therefore, *XIST* is an appealing solu-

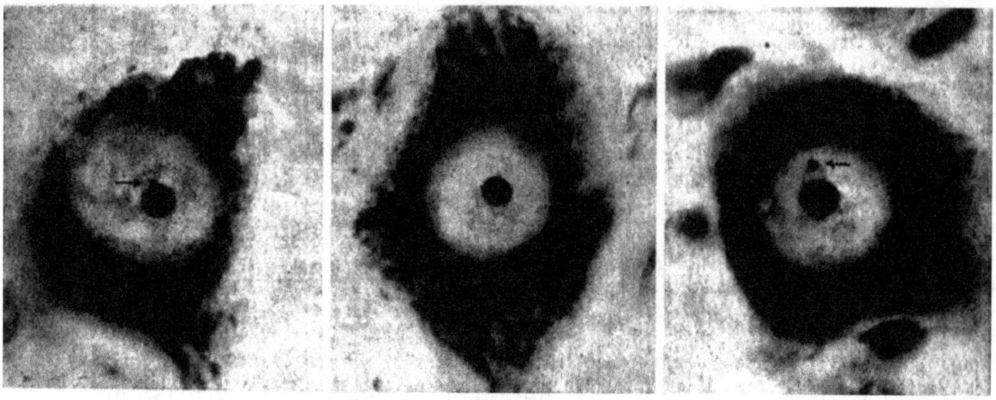

Fig. 4.5 The inactive X chromosome observed in neurons of female cats. Microscopy images of stained cat neurons show the cell darkly and the nucleus lightly stained. Within the nucleus, the large dark round spot is the nucleolus. Near the nucleolus, a smaller dark spot (arrow) can be observed in female neurons (left and right) but not in those of males (center). This is the inactive X chromosome which, in recognition of its discoverer, is also referred to as Barr body. (From Barr and Bertram (1949))

tion to restrict dosage compensation to one of the two X chromosomes in female cells. *XIST* specifically associates with the chromosome from which it is produced. The X chromosome that does not express *XIST* remains active. Since male cells contain only a single active X chromosome, they do not express *XIST* in general. Conceptually, the issue arises as to how *XIST* prevents its own repression on the Xi, when the majority of gene promoters are inactivated. Actually, a number of genes are known that escape X inactivation and remain expressed from the Xi. This observation points to the existence of specific expression control elements in the promoters of *XIST* and genes that escape X inactivation, but this aspect remains incompletely understood.

XIST localization to the Xi induces modifications of chromatin and gene repression. The formation of a fully inactivated X chromosome then occurs through sequential changes to Xi. Studies in mouse embryonic stem cells (ESCs), which recapitulate the process of X chromosome inactivation, have led to the view of a multistep process. *Xist* RNA recruits chromatin modifiers including Polycomb group (PcG) proteins and the mammalian homologue of *Split ends* (SPEN), an RNA-binding protein. Initially, repression and chromatin changes are dependent on *XIST* expression. Reactivation and complete reversion to an active X chromosome is observed when *Xist* is lost at this stage. This phase of X inactivation is referred to as the initiation phase, when *Xist*-dependent gene repression and chromatin changes are established that are reversible (Fig. 4.6a). The initiation phase is distinguished from the maintenance of gene repression on Xi in differentiated somatic cells. During development, the Xi acquires a series of chromatin modifications as cells differentiate into mature cell types of the tissues. Chromatin changes on the Xi include the deposition of the histone H2A variant macroH2A, the loss of acetylation of histone H4, and the acquisition of DNA methylation at promoters of repressed genes. At this stage, *Xist* becomes dispensable for maintaining gene repression on the Xi. Therefore, the repressed state is perpetuated in a heritable manner without *Xist*. Stable maintenance of gene repression that is independent of *XIST* characterizes the maintenance phase of X chromosome inactivation (Fig. 4.6a). The fact that *Xist*,

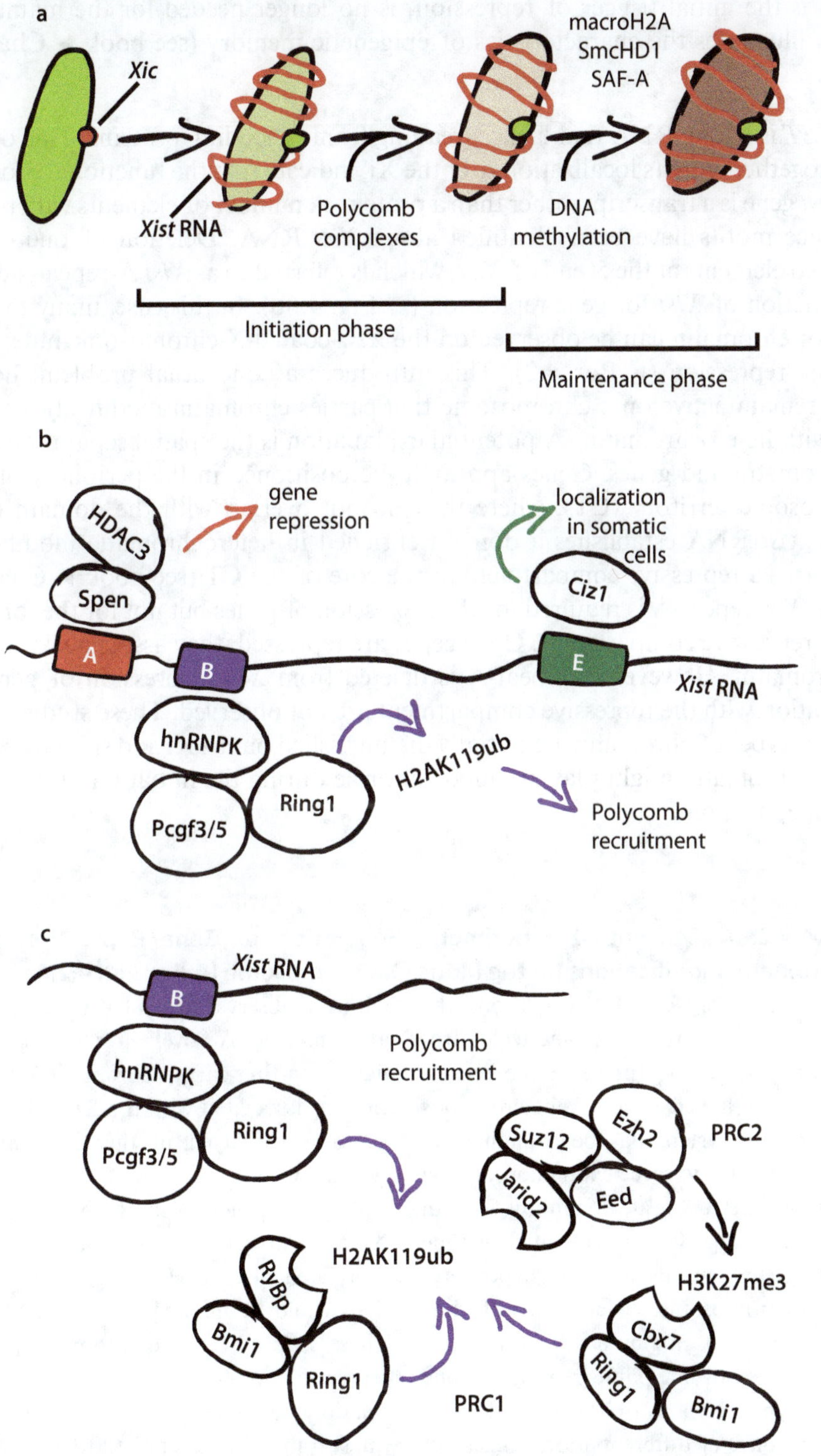

Fig. 4.6 Chromosome-wide gene silencing during X chromosome inactivation. **a** Developmental changes of chromatin on the Xi are aligned with the initiation and maintenance phase of X chromosome inactivation. **b** Molecular interaction of *Xist* A-, B-, and E-repeats are indicated. **c** Recruitment of PRC1 and PRC2 in X chromosome inactivation is shown. These pathways lead to chromosome-wide modification of the Xi with H3K27me3 and H2AK119ub

which is the initial trigger of repression, is no longer needed for the maintenance phase, illustrates the characteristics of epigenetic memory (see book ▶ Chap. 3 of Paro).

XIST is a long RNA that lacks a reading frame encoding proteins. This observation, together with its localization over the Xi, indicate that the functional product of the *Xist* gene is a transcript rather than a protein. A number of elements with repeated sequence motifs have been identified along *Xist* RNA. Deletion of one of these repeated elements at the 5′ end of *Xist*, which is referred to as *Xist* A-repeat, abolishes the function of *Xist* for gene repression (◘ Fig. 4.6b). In this case, many modifications of chromatin can be observed on the *Xist*-coated X chromosome independent of gene repression (▶ Box 4.2). This introduces a conceptual problem: how can genes remain active on a chromosome that carries chromatin modifications associated with heterochromatin? A potential explanation is the spatial separation of heterochromatin and genes. Genes appear to be positioned in the periphery of the X chromosome territory (CT), where they do not overlap with the domain of *Xist* RNA. *Xist* RNA establishes a domain enriched in heterochromatic modifications and forms a repressive compartment in the core of the CT (see book ▶ Chap. 1 of Wutz). *Xist* repeat-A is required for the repression of genes but not for the formation of the repressive compartment. Once genes are repressed, they associate with Xi heterochromatin. However, if repeat-A is deleted from *Xist*, repression of genes and association with the repressive compartment are not observed. These studies suggest that two types of chromatin are initially distinguished on the Xi and spatial organization of chromatin might play an important role during the initiation of X chromosome inactivation.

Box 4.2: A Conceptual Experiment: Separation of Gene Repression and Chromatin Modifications on the Mouse X Chromosome (◘ Box Fig. 4.2)

Expression of *Xist* RNA without the A-repeat leads to chromosome-wide chromatin changes but prevents the repression of X-linked genes. A heterologous expression system has been engineered in male ESCs to force expression of *Xist* without the A-repeat. **a** Staining with an antiserum specific for acetylated histone H4 (green) reveals a single hypoacetylated chromosome, which is identified as the X chromosome by DNA FISH using an X chromosome paint (red). **b** Histone H4 hypoacetylation is observed even after 9 days of differentiation, consistent with the observation that X-linked genes are expressed. A novel chromosome configuration that carries chromatin modifications associated with the Xi and active genes is displayed in this experiment. A magnified view of this Xiag (for Xi with active genes) is shown on the right. Notably, stripes of chromatin containing acetylated histone H4 can be discerned that would be missing from a silent Xi. These stripes of active chromatin might represent clusters of active genes that are spatially separated from heterochromatin. (From Pullirsch et al. (2010))

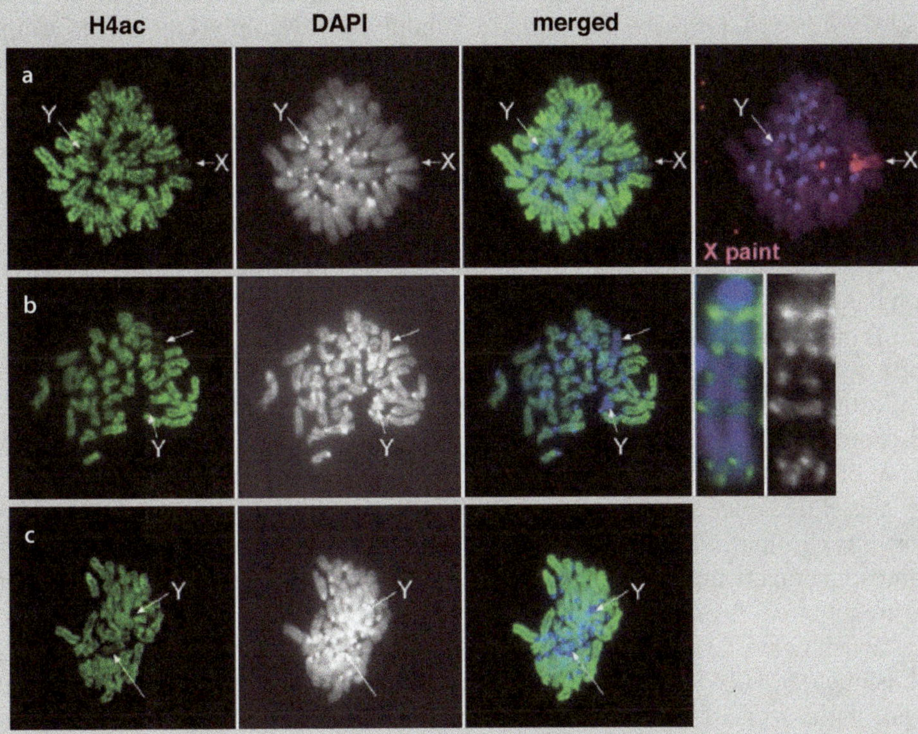

◘ Box Fig. 4.2 Chromosome-wide histone H4 hypoacetylation is triggered by Xist independent of gene silencing

The mechanism of gene repression by *Xist* has been investigated. The RNA-binding protein SPEN binds to the *Xist* A-repeat (◘ Fig. 4.6b) and is required for the function of *Xist* in gene repression. Mutations affecting SPEN phenocopy the deletion of *Xist* repeat-A in cultured mouse ESCs. SPEN associates with NCoR/SMRT co-repressor complexes, which contain the histone deacetylase HDAC3. The latter suggests that histone deacetylation contributes to the gene silencing pathway of *Xist*. A conceptually important question is how *Xist* function in repressing genes can be localized to the X chromosome from which it is expressed. The RNA binding protein hnRNPU is required for localizing *Xist* RNA to Xi in specific cellular contexts. Notably, hnRNPU is a protein that associates with the nuclear scaffold and is also referred to as Scaffold attachment factor A (SAF-A). Interactions with non-chromatin nuclear organization seems to contribute to the specific localization of *Xist* to the Xi. Recently, the cyclin binding protein CIZ1 has also been shown to be required for *Xist* localization in differentiated somatic cells but not in embryonic cells. CIZ1 is enriched at the Xi, apparently mediated by the *Xist* repeat-E (◘ Fig. 4.6b).

During the initiation phase of X chromosome inactivation, PcG proteins (see book ► Chap. 3 of Paro) are recruited to the Xi. This *Xist* RNA-dependent recruitment mechanism has been analyzed in detail. The RNA-binding protein hnRNPK is required for PcG recruitment to the Xi. *Xist* repeat-B and repeat-C have been shown to represent binding sites for hnRNPK. Initially, a complex containing the PcG proteins PCGF3 or PCGF5 and the catalytic subunit of Polycomb Repressive Complex 1 (PRC1), RING1B (◘ Fig. 4.6b), is recruited by hnRNPK to *Xist*. This atypical PRC1 complex mediates ubiquitinylation of histone H2A lysine 119 (H2AK119ub). This modification has two functions. Firstly, it can recruit additional PRC1 complexes that contain RyBb which has affinity for H2AK119ub. Secondly, Jarid2 binds H2AK119ub and activates PRC2. PRC2 contains the histone methyltransferase EZH2 and catalyzes tri-methylation of histone H3 lysine 27 (H3K27me3) on the Xi. H3K27me3 can, in turn, enhance the recruitment of PRC2 through the EED protein, which has affinity for H3K27me3. In addition, recruitment of canonical PRC1 complexes is mediated through the CBX7 subunit, which contains a chromodomain with affinity for H3K27me3. Multiple interactions between PcG proteins and histone modifications generate a feed-forward loop that results in a strong enrichment of PcG proteins over Xi (◘ Fig. 4.6b). Spreading of PcG complexes leads to a high amount of histone modifying activity and, therefore, nearly all histone H3 on the Xi carry the H3K27me3 modification. Multiple interactions between PcG complexes endow this chromatin modification system with a substantial robustness. Random X chromosome inactivation appears not materially affected when individual PcG proteins are mutated. A mutation in the EED gene diminishes PRC2 activity and blocks the establishment of H3K27me3 on Xi but not the recruitment of PRC1 complexes, establishment of H2AK119ub, and gene repression. Furthermore, additional requirements for the recruitment of PcG proteins to Xi appear to exist. For example, a depletion of SPEN has been shown to reduce the amount of PcG complexes that are recruited by *Xist*. However, it is not clear whether SPEN directly interacts with PcG complexes or indirectly causes an effect on PcG recruitment by influencing gene repression or *Xist* RNA modification. It is noteworthy that a deletion of *Xist* repeat-A appears to have less effect on PcG recruitment by *Xist* than the mutation of SPEN. Future studies will be required to explain these observations.

Additional factors that contribute to the initiation of X chromosome inactivation include lamin B receptor (LBR) and a complex that mediates methylation of adenines on RNA. It is likely that several of these molecular pathways act in parallel and, through their combination, bring about the remarkable localization and repressive function of *Xist* RNA. Conceptually, it is tempting to speculate that *Xist* RNA might contain multiple binding sites for different complexes, repeat-A, -B and –C, and –E are examples in support of this view (◘ Fig. 4.6b). Thus, X inactivation might instantiate novel combinatorial functions of molecular components that have other roles in distinct biological processes.

The maintenance of X chromosome inactivation involves other factors than its initiation. The atypical structural maintenance of chromosomes (SMC) protein SMC hinge domain 1 (SmcHD1) has been identified in a genetic screen in mice. A mutation in *SmcHD1* causes female-specific lethality, suggesting a role in dosage compensation (Blewitt et al. 2008). Subsequent studies could show that *SmcHD1* is required for the maintenance of gene repression and promoter DNA methylation in somatic cells. In contrast, the initiation of X chromosome inactivation is largely unaffected by a mutation in *SmcHD1*. *SmcHD1* is conceptually important as it is a rare example of a gene whose mutation causes sex-specific lethality in mammals (another one being *Xist*). Many components of the dosage compensation system have additional functions that are also required in males. For example, the *hnRNPK* mutation is embryonic lethal in mice of both sexes (Gallardo et al. 2015).

Although most of the genes on the Xi are transcriptionally silenced, certain genes remain active on the Xi. Some of these genes are located in regions of the X chromosome that are homologous with the Y chromosome and exchange genetic information through recombination during meiosis in male germ cells through X-Y pairing. Genes within these pseudoautosomal regions (PARs) of the X chromosome are present in two copies in both sexes, similar to autosomal genes. Therefore, these genes do not require dosage compensation and remain transcriptionally active on the Xi. In addition, several genes that do not possess homologues on the Y chromosome are expressed from the Xi. These genes are referred to as escape genes and the phenomenon is called "escape from X inactivation". Escape genes are important for understanding phenotypes that arise from numerical aberrations of sex chromosomes including the human Klinefelter (XXY) and Turner (XO) syndromes. The level of expression of escape genes from the Xi often does not reach that of the X chromosome that remains active (Xa). Furthermore, for some genes, tissue-specific escape or polymorphic escape among individuals has been observed (Carrel and Willard 1999, 2005). The molecular features that determine whether a gene is repressed or remains active on the Xi remain to be explored further. Notably, it appears that a higher frequency of escape genes can be observed in regions of the X chromosome that were recently acquired from autosomal fusions. This correlation could indicate that chromosomal features change in evolution to accommodate dosage compensation.

4.3.2 Regulation of XCI in Different Mammals

The *Xist* gene originated before the radiation of placental mammals[3] (◘ Fig. 4.1). However, mammalian sex chromosomes and dosage compensation systems have continued to evolve after the dosage compensation system was established. Even today, evolution continues to introduce changes into sex chromosome and dosage compensation systems among placental mammals, which can be illustrated by striking examples. Dominant mutations that specify female fate have arisen on the X chromosomes of lemmings and African pygmy mice (Veyrunes et al. 2010). As a consequence, females that possess one X chromosome carrying a dominant modifier and a Y chromosome (X*Y females) can be observed along with XX and XX* females, and XY males. Possibly, the increased ratio of females over males was favored in these species. In contrast, novel autosomal male sex determining factors can be observed in specific vole species (Mulugeta et al. 2016). A novel autosomal sex determination mechanism has led to the loss of the second sex chromosome in voles of the species *Ellobius lutescens*. A single X chromosome is present in both males and females and likely X chromosome inactivation is no longer initiated. *E. talpinus* and *E. tancrei* possess two X chromosomes in both sexes and have lost their Y chromosomes entirely. It is likely that X chromosome inactivation occurs in both sexes of

3 Evolutionary radiation of placental mammals followed the extinction of the dinosaurs 65 million years ago. In the late Cretaceous, placental mammals were mostly represented by small shrew-like animals from which such diverse species as mice, bats, dolfins, cows, dogs, and humans evolved.

these species. These examples illustrate dramatic changes in sex determination strategies between closely related species. They are accompanied by a gain or loss of heteromorphic sex chromosomes and dosage compensation processes. The presence of an inactive X chromosome in *E. talpinus* males and females bears some reminiscence of the modified fourth chromosome in *D. melanogaster* as an evolutionary relict of a past sex chromosome system.

In placental mammals, chromosome-wide silencing is a consequence of *Xist* expression. The regulation of *Xist* is, therefore, important for correct dosage compensation whereby, in males, a single X chromosome remains active and, in females, one of the two X chromosomes is inactivated. Conceptually, it has been proposed that blocking factors would prevent *Xist* activation in male cells. In female cells, X-linked activation factors would be produced from two X chromosomes in sufficient amounts to counteract the blocking factors and lead to *Xist* expression. Blocking factors are expected to be encoded on autosomes and to set a threshold in male and female cells for X-linked activation factors to induce *Xist* expression. The identity of these blocking factors and X-linked activators remains to be fully understood but several genes with predicted properties have been identified (◘ Fig. 4.7a). These genes appear to be species-specific, likely reflecting the evolutionary divergence of placental mammals.

In mice, the X-linked gene *Rnf12/RLim* can trigger *Xist* expression in male ESCs when it is expressed from transgenes. This observation suggests that *Rnf12/RLim* is an X-linked activator of X chromosome inactivation. The function of *Rnf12/RLim* in female mouse embryos appears to be restricted to the initiation of imprinted X inactivation in early embryos (◘ Fig. 4.7b). Therefore, it has been proposed that other activators might act in the regulation of random X chromosome inactivation in mice.

Also, the mechanism of blocking factor function has been studied. A potential gene that likely acts downstream of blocking factors and negatively regulates *Xist* expression is the non-coding RNA gene *Tsix*, whose name is derived from the reverse spelling of *Xist* because it is transcribed in antisense orientation to *Xist* (◘ Fig. 4.7a). *Tsix* represses the promoter of *Xist* on the same chromosome from which it is expressed and, therefore, can be considered a *cis*-acting repressor. *Tsix* is the target of autosomal factors that enhance *Tsix* transcription and can be considered as blocking factors (◘ Fig. 4.7a). Among these factors are several transcription factors that are expressed in early embryos and repress *Xist*. These factors are not exclusively functioning in dosage compensation but have also other roles in embryo development.

For understanding the regulation by blocking factors, it is helpful to examine the function of the *Tsix* and *XACT* genes in mice and humans, respectively. Both genes are X-linked and produce non-coding transcripts. They are *cis*-acting repressors of *Xist* and their expression is associated with Xa. Yet, *Tsix* and *XACT* are not conserved between mouse and human and likely act by different molecular mechanisms. *XACT* encodes a non-coding transcript that associates with the chromosome territory of the Xa and appears to counteract *XIST* localization and gene repression. In humans, both X chromosomes in female cells initially express *XIST* but X-linked gene silencing appears to be incomplete. The reduced gene expression caused by this early wave of *XIST* expression is referred to as "dampening" instead of silencing of X-linked gene expression and is exclusively observed in humans. It is conceivable that antagonistic interactions between *XACT* and *XIST* contribute to dampening. As

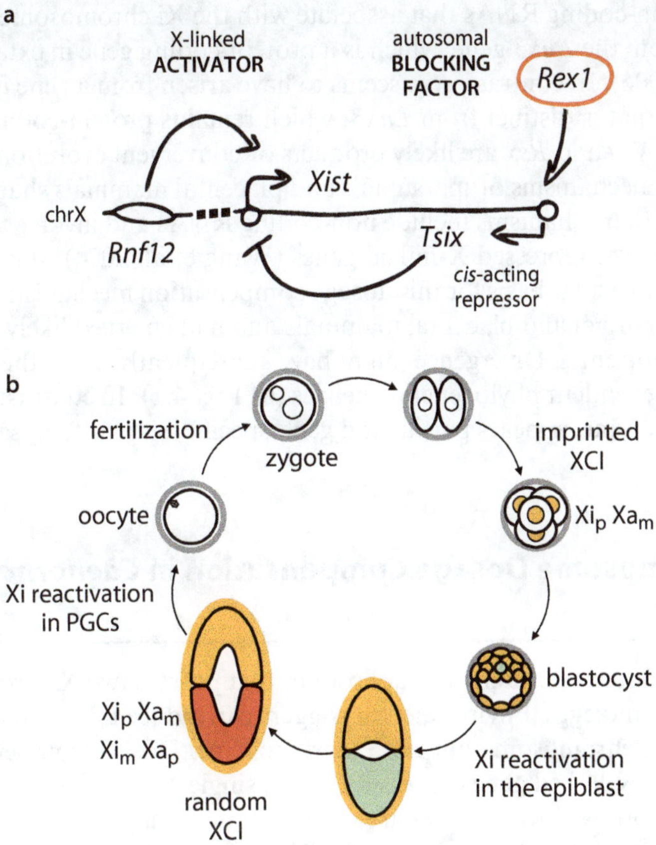

Fig. 4.7 Regulation of X chromosome inactivation in mice. **a** *Xist* expression is stimulated by X-linked activators, including *Rnf12*. Blocking factors enhance expression of *Tsix* that is a *cis*-acting repressor of *Xist*. In males, X-linked activators are not abundant enough to overcome the repressive effect of blocking factors and X inactivation is not initiated. **b** Imprinted (orange) and random (red) X chromosome inactivation in mouse embryos. Initially, the paternally inherited X chromosome in inactivated in 4-cell embryos. The Xi is maintained in the extraembryonic membranes and placenta but reactivated in the epiblast. After implantation, random inactivation of the maternal or paternal X chromosome is initialed in the cells of the developing embryo. Once the Xi is chosen, all descendants of the cell maintain this choice, giving rise to a genetic mosaic of patches of cells

embryo development progresses, one X chromosome maintains *XIST* expression and initiates gene silencing whereas the second X chromosome remains active and is covered by *XACT*. *XACT* expression ceases as cell differentiation progresses further, suggesting a role for *XACT* specifically in regulating the initiation phase of X inactivation. The observation that independent *cis*-acting repressors of *Xist* have evolved in mice and humans suggests a model where blocking factors activate these X-linked RNAs, thereby counteracting the initiation of X inactivation. If this view were correct, blocking factor abundance or activity would be integrated at the promoters of *Tsix* and *XACT*.

In marsupials, inactivation of the paternally inherited X chromosome is observed. Imprinted X inactivation is distinct from random X inactivation that is exclusive to placental mammals (◘ Fig. 4.1). In the American opossum *Monodelphis domestica*, *Rsx* appears to possess a similar function as *Xist* in mice (Grant et al. 2012). *Rsx* and

Xist encode non-coding RNAs that associate with the Xi chromosome territory. *Xist* has evolved from the *Lnx3* gene, which is a protein-coding gene in extant vertebrates (Duret et al. 2006). In contrast, *Rsx* seems to have arisen from a gene in proximity to the *Hprt* gene that is distinct from *Lnx3*, which remains protein-coding in marsupials. Therefore, *Xist* and *Rsx* are likely products of convergent evolution as the dosage compensation mechanisms of marsupials and placental mammals share a number of similarities. Both mechanisms include non-coding RNAs and involve the acquisition of H3K27me3 over repressed X-linked genes (Wang et al. 2014). It could be speculated that the molecular basis for this dosage compensation mechanism existed before the split of marsupial and placental mammals and had emerged likely soon after the split from monotremes. Divergence might have subsequently led to the origin of *Xist* and *Rsx* in independent phylogenetic lineages (◘ Fig. 4.1). In contrast, dosage compensation in platypus appears partial and gene-specific, similar to dosage compensation in birds.

4.4 X Chromosome Dosage Compensation in *Caenorhabditis elegans*

C. elegans reproduces through hermaphrodites that possess two X chromosomes and produce sperm and eggs. Environmental triggers can induce the production of males with a single X chromosome but no Y chromosome. Dosage compensation in the worm equalizes X-linked gene expression to the single X chromosome in males. For this, genes on both X chromosomes are repressed in hermaphrodites. *C. elegans* possesses a dosage compensation complex (DCC) that contains structural maintenance of chromosomes (SMC) proteins and resembles a condensin-like complex (see book ► Chap. 1 of Wutz). Therefore, the composition of the fly and worm DCCs are different and, in the worm, the DCC is recruited to both X chromosomes of hermaphrodites. DCC recruitment leads to an impairment of transcription to reduce the gene expression level by half to match that of the single male X.

Components of the *C. elegans* DCC have been identified by screening for sex-specific lethal mutations. Whereas defects in dosage compensation cause male lethality in *D. melanogaster*, they lead to lethality of hermaphrodites in *C. elegans*. The SMC protein MIX-1, together with several subunits that are shared with condensin, make up the DCC (◘ Fig. 4.8). The SMC family of proteins has functions in organizing chromatin during cell division, coordination of gene expression, and DNA repair. In the case of the *C. elegans* DCC, SMC proteins mediate a partial repression of transcription on both X chromosomes. As a consequence of DCC localization, H4K20me1 is enriched on the two X chromosomes.

In *C. elegans*, the number of X chromosomes determines sex and initiates dosage compensation (◘ Fig. 4.8). The DCC is engaged by SDC-2, which also functions to suppresses male development. SDC-2 localizes to both X chromosomes together with the DCC. The dosage compensated X chromosomes contain low levels of H4K16ac. Recent studies provided evidence that deacetylation of H4K16ac can already be observed early in worm development before the DCC is recruited. This early deacetylation seems to require a PRC2-like complex that contains MES2, a homologue of E(z), MES3, and MES6. This suggests the presence of an initiation

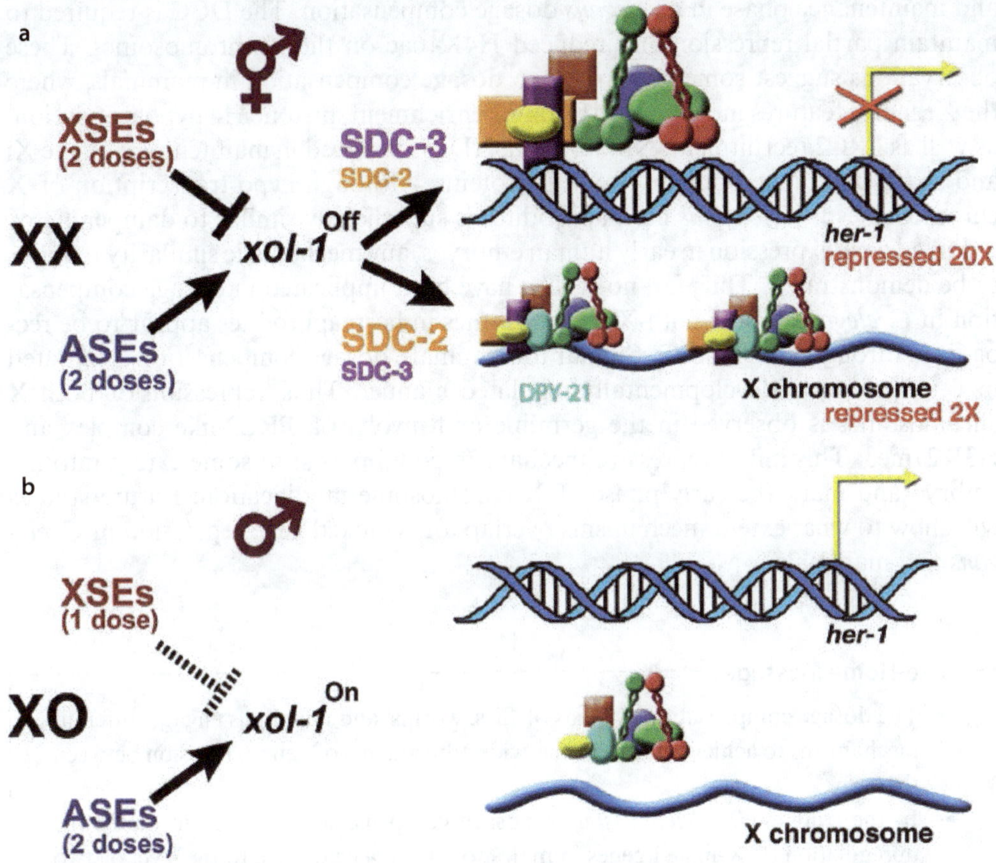

Fig. 4.8 Genetic control of sex determination and dosage compensation in *C. elegans*. **a** *xol-1* is the master sex-determination gene that controls both sex determination and dosage compensation. It is the direct molecular target of the X-chromosome counting mechanism that determines sex. In XX hermaphrodites, two doses of the X-signal elements (XSEs) repress *xol-1* by overcoming the *xol-1* activation achieved by autosomal signal elements (ASEs). *xol-1* repression occurs at the level of transcription and pre-mRNA splicing. When *xol-1* is repressed, the XX-specific gene *SDC-2* is active and stabilizes SDC-3. SDC-2 acts with SDC-3 to target dosage compensation proteins to the X chromosome, thereby repressing gene expression by half. SDC-2 plays the lead role in recognizing X-specific sequences. It is the only dosage compensation protein expressed solely in XX animals. SDC-2 also activates the hermaphrodite program of sexual development by repressing the male-specific sex-determination gene *her-1*. SDC-2 acts with SDC-3 to recruit dosage compensation proteins to *her-1*. In this case, SDC-3 plays the lead role in recognizing *her-1* DNA targets. The *her-1* and X complexes differ by one component: DPY-21 is present on the X but not at *her-1*. **b** In XO males, the single dose of XSEs is insufficient to overcome the activating influence of ASEs. *xol-1* is active and promotes male fate by repressing the activities of *sdc* genes. *her-1* is transcribed and the single X is not repressed. In *xol-1* XO mutants, the dosage compensation complex localizes to the single X chromosome, killing XO animals by reducing X-linked gene expression. (Copyright: © 2005 Barbara J. Meyer. From Meyer, B. J. X-Chromosome dosage compensation (June 25, 2005), WormBook, ed. The *C. elegans* Research Community, WormBook, doi/► https://doi.org/10.1895/wormbook.1.8.1, ► http://www.wormbook.org., which permits unrestricted use, distribution, and reproduction in any medium, provided the original author and source are credited)

and maintenance phase in *C. elegans* dosage compensation. The DCC is required to maintain partial repression and reduced H4K16ac on the X chromosomes. These observations suggest some similarities to dosage compensation in mammals, where the Xi shares features including H4K20me1 enrichment, histone H4 hypoacetylation, as well as PRC2 recruitment. Notably, SmcHD1 is involved in maintenance of the Xi and shares the hinge domain of SMC proteins. Although hypo-transcription of X chromosomes in *C. elegans* hermaphrodites is superficially similar to dampening of X-linked gene expression in early human embryos, any mechanistic similarity remains to be demonstrated. Thus far, no RNAs have been implicated in dosage compensation in *C. elegans*, where both X chromosomes in hermaphrodites appear to be recognized through DNA motifs. Similar to mammals, dosage compensation is initiated in *C. elegans* in a developmentally regulated manner. Thus, repression of both X chromosomes is observed in the germline and involves a PRC2-like complex and H3K27me3. This initial repressive mechanism could persist to some extent into the embryo and mark the early phase of X chromosome modification. Future studies will show to what extent mechanisms overlap for X-linked gene repression in *C. elegans* and mammals.

Take-Home Message

- The dosage compensation systems of flies, worms, and mammals engage different mechanisms to achieve chromosome-wide adjustments of gene expression between the sexes.
- In the fruit fly *D. melanogaster*, a dosage compensation complex mediates the upregulation of X-linked genes in males to adjust gene dosage to the level of two X chromosomes as is observed in females.
- In mammals, the process of X chromosome inactivation silences one of the two X chromosomes in female cells, thereby equalizing the gene dose to the single X chromosome of males.
- In the nematode *C. elegans*, the two X chromosomes in hermaphrodites are partially repressed and adjust the gene dose to a level similar to males that possess a single X chromosome.
- Components of dosage compensation systems have functions in animal development in addition to dosage compensation and can, therefore, be considered model systems for understanding gene regulation.
- The fly and mammalian dosage compensation systems include long non-coding RNAs (*Xist*, *Rsx*, *XACT*, *roX1*, and *roX2*) as part of the mechanism.
- Chromosome-wide dosage compensation mechanisms have, thus far, been discovered in XY systems in a few animal species, although sex chromosome systems are widely distributed among animals and in dioecious plants (Vyskot and Hobza 2015). It is possible that novel molecular reagents facilitate further discoveries. Considering past discoveries made in dosage compensation systems, future research in this area promises to extend our understanding of the molecular mechanisms and the evolution of pathways for gene regulation.

References

Bachtrog D, Mank JE, Peichel CL, Kirkpatrick M, Otto SP, Ashman TL, Hahn MW, Kitano J, Mayrose I, Ming R, Perrin N, Ross L, Valenzuela N, Vamosi JC, Tree of Sex C (2014) Sex determination: why so many ways of doing it? PLoS Biol 12(7):e1001899. https://doi.org/10.1371/journal.pbio.1001899

Barr ML, Bertram EG (1949) A morphological distinction between neurones of the male and female, and the behaviour of the nucleolar satellite during accelerated nucleoprotein synthesis. Nature 163(4148):676

Bellott DW, Skaletsky H, Pyntikova T, Mardis ER, Graves T, Kremitzki C, Brown LG, Rozen S, Warren WC, Wilson RK, Page DC (2010) Convergent evolution of chicken Z and human X chromosomes by expansion and gene acquisition. Nature 466(7306):612–616. https://doi.org/10.1038/nature09172

Blewitt ME, Gendrel AV, Pang Z, Sparrow DB, Whitelaw N, Craig JM, Apedaile A, Hilton DJ, Dunwoodie SL, Brockdorff N, Kay GF, Whitelaw E (2008) SmcHD1, containing a structural-maintenance-of-chromosomes hinge domain, has a critical role in X inactivation. Nat Genet 40(5):663–669. https://doi.org/10.1038/ng.142

Bone JR, Lavender J, Richman R, Palmer MJ, Turner BM, Kuroda MI (1994) Acetylated histone H4 on the male X chromosome is associated with dosage compensation in Drosophila. Genes Dev 8(1):96–104

Carrel L, Willard HF (1999) Heterogeneous gene expression from the inactive X chromosome: an X-linked gene that escapes X inactivation in some human cell lines but is inactivated in others. Proc Natl Acad Sci U S A 96(13):7364–7369

Carrel L, Willard HF (2005) X-inactivation profile reveals extensive variability in X-linked gene expression in females. Nature 434(7031):400–404. https://doi.org/10.1038/nature03479

Duret L, Chureau C, Samain S, Weissenbach J, Avner P (2006) The Xist RNA gene evolved in eutherians by pseudogenization of a protein-coding gene. Science 312(5780):1653–1655. https://doi.org/10.1126/science.1126316

Gallardo M, Lee HJ, Zhang X, Bueso-Ramos C, Pageon LR, McArthur M, Multani A, Nazha A, Manshouri T, Parker-Thornburg J, Rapado I, Quintas-Cardama A, Kornblau SM, Martinez-Lopez J, Post SM (2015) hnRNP K is a haploinsufficient tumor suppressor that regulates proliferation and differentiation programs in hematologic malignancies. Cancer Cell 28(4):486–499. https://doi.org/10.1016/j.ccell.2015.09.001

Grant J, Mahadevaiah SK, Khil P, Sangrithi MN, Royo H, Duckworth J, McCarrey JR, VandeBerg JL, Renfree MB, Taylor W, Elgar G, Camerini-Otero RD, Gilchrist MJ, Turner JM (2012) Rsx is a metatherian RNA with Xist-like properties in X-chromosome inactivation. Nature 487(7406):254–258. https://doi.org/10.1038/nature11171

Graves JA (2015) Weird mammals provide insights into the evolution of mammalian sex chromosomes and dosage compensation. J Genet 94(4):567–574

Johansson AM, Stenberg P, Bernhardsson C, Larsson J (2007) Painting of fourth and chromosome-wide regulation of the 4th chromosome in Drosophila melanogaster. EMBO J 26(9):2307–2316. https://doi.org/10.1038/sj.emboj.7601604

Kaiser VB, Bachtrog D (2010) Evolution of sex chromosomes in insects. Annu Rev Genet 44:91–112. https://doi.org/10.1146/annurev-genet-102209-163600

Marshall Graves JA (2008) Weird animal genomes and the evolution of vertebrate sex and sex chromosomes. Annu Rev Genet 42:565–586. https://doi.org/10.1146/annurev.genet.42.110807.091714

Mulugeta E, Wassenaar E, Sleddens-Linkels E, van IJcken WFJ, Heard E, Grootegoed JA, Just W, Gribnau J, Baarends WM (2016) Genomes of Ellobius species provide insight into the evolutionary dynamics of mammalian sex chromosomes. Genome Res 26(9):1202–1210. https://doi.org/10.1101/gr.201665.115

Pullirsch D, Hartel R, Kishimoto H, Leeb M, Steiner G, Wutz A (2010) The Trithorax group protein Ash2l and Saf-A are recruited to the inactive X chromosome at the onset of stable X inactivation. Development 137(6):935–943. https://doi.org/10.1242/dev.035956

Ruiz MF, Esteban MR, Donoro C, Goday C, Sanchez L (2000) Evolution of dosage compensation in Diptera: the gene maleless implements dosage compensation in Drosophila (Brachycera suborder) but its homolog in Sciara (Nematocera suborder) appears to play no role in dosage compensation. Genetics 156(4):1853–1865

Stenberg P, Larsson J (2011) Buffering and the evolution of chromosome-wide gene regulation. Chromosoma 120(3):213–225. https://doi.org/10.1007/s00412-011-0319-8

Veyrunes F, Waters PD, Miethke P, Rens W, McMillan D, Alsop AE, Grutzner F, Deakin JE, Whittington CM, Schatzkamer K, Kremitzki CL, Graves T, Ferguson-Smith MA, Warren W, Marshall Graves JA (2008) Bird-like sex chromosomes of platypus imply recent origin of mammal sex chromosomes. Genome Res 18(6):965–973. https://doi.org/10.1101/gr.7101908

Veyrunes F, Chevret P, Catalan J, Castiglia R, Watson J, Dobigny G, Robinson TJ, Britton-Davidian J (2010) A novel sex determination system in a close relative of the house mouse. Proc Biol Sci 277(1684):1049–1056. https://doi.org/10.1098/rspb.2009.1925

Vicoso B, Bachtrog D (2013) Reversal of an ancient sex chromosome to an autosome in Drosophila. Nature 499(7458):332–335. https://doi.org/10.1038/nature12235

Vyskot B, Hobza R (2015) The genomics of plant sex chromosomes. Plant Sci 236:126–135. https://doi.org/10.1016/j.plantsci.2015.03.019

Wang X, Douglas KC, Vandeberg JL, Clark AG, Samollow PB (2014) Chromosome-wide profiling of X-chromosome inactivation and epigenetic states in fetal brain and placenta of the opossum, Monodelphis domestica. Genome Res 24(1):70–83. https://doi.org/10.1101/gr.161919.113

Zhang Y, Malone JH, Powell SK, Periwal V, Spana E, Macalpine DM, Oliver B (2010) Expression in aneuploid Drosophila S2 cells. PLoS Biol 8(2):e1000320. https://doi.org/10.1371/journal.pbio.1000320

5

Genomic Imprinting: An Epigenetic Phenomenon

Contents

What You Will Learn in This Chapter

A typical cell contains two sets of chromosomes: one that was inherited from the mother, the other from the father. Usually, autosomal alleles are expressed at similar levels from the maternally and paternally inherited chromosomes. This chapter is dedicated to an exception of this rule: the expression of genes that are regulated by genomic imprinting depends on the parental origin of the allele, leading to the non-equivalence of maternal and paternal genomes. Genomic imprinting is a paradigm of epigenetic gene regulation as genetically identical alleles can exist in two expression states within the same nucleus. The imprints marking the parental alleles are established in the parental germline, maintained during the development of the offspring, but reset before they are passed on to the next generation. In mammals, the primary imprint is usually a differentially methylated region at the locus but there are also examples where histone modifications mark the parental alleles. Many imprinted genes play important roles for development and are associated with human disease. Interestingly, genomic imprinting evolved independently in mammals and seed plants and similar mechanisms have been recruited to regulate imprinted expression in the two kingdoms. We will discuss evolutionary constraints that could have led to the evolution of genomic imprinting in these seemingly disparate lineages.

5.1 Discovery of the Non-equivalence of Maternal and Paternal Genomes

In general, the genetic contributions from father and mother are equivalent (except for the sex chromosomes; see book ► Chap. 4 of Wutz), each contributing a chromosome set to the progeny. Genetically, this manifests itself by the recessive nature of the vast majority of loss-of-function mutations, i.e., if a mutant allele is only inherited from one of the parents, the offspring is phenotypically wild-type. Studies on traits that did show surprising parent-of-origin-specific effects suggested the non-equivalence of the parental genomes and led to the discovery of genomic imprinting.

5.1.1 Genome-Wide Imprinting in Insects

The first evidence that parental contributions may not be equivalent came from cytogenetic observations that Charles Metz made nearly 100 years ago. He found that, during the first meiotic division in the male germline of the fungus gnat *Sciara coprophila*, an entire chromosome set was eliminated. Genetic studies using the autosomal recessive *truncate* mutant, which affects wing shape, showed that only the maternal allele was transmitted by the sperm while the paternal chromosomes were lost. Metz concluded (Metz 1938): "… that the difference in behavior [of the chromosomes] is due to a general qualitative difference between maternal and paternal chromosomes …, and that this [difference] is impressed on the chromosomes in the preceding generation by the sex of the parent." And he continued: "It should be emphasized that this qualitative difference or modification persists for only one generation and is reversible." Although the term "imprint" was only introduced by Helen Crouse when studying the genetic determinants of paternal X-chromosome elimination in *S. coprophila*, Metz' conclusions set the foundations of the imprinting field, namely:

1. The differential behavior of the chromosomes implies the existence of differential modifications, i.e., imprints, on the chromosomes.
2. Imprints may be on the maternal or paternal chromosomes, or both; the imprint represents a difference between them, not an absolute state.
3. Imprints are established in the parents but persist during development of the progeny. Thus, they must be maintained during cell division as the genetic material is replicated.
4. Imprints are reversible and have to be reset in every generation: both maternal and paternal chromosome sets of a male are passed on to the progeny with paternal imprints, those of a female with maternal imprints.

Parental imprints were later also postulated in coccids, scale insects where the paternal genome is heterochromatinized and silenced or even eliminated early during embryogenesis. Although maternal and paternal chromosome sets in these embryos are distinctly marked (e.g., by differences in DNA methylation, H3K9me3, and H4ac), the mechanisms underlying the establishment, maintenance, and resetting of the imprints are not well understood (Sánchez 2014). While studies in insects provided a conceptual framework for imprints that distinguish maternal and paternal chromosomes, the focus of this chapter is not on genome-wide imprinting but rather on the imprints that distinguish individual genes or gene clusters and ultimately lead to the parent-of-origin-dependent expression of the corresponding alleles.

5.1.2 Discovery of Genomic Imprinting at an Individual Locus in Maize

Genomic imprinting regulating an individual gene rather than an entire chromosome set was first discovered when studying seed coloration in maize (*Zea mays*). In flowering plants, seeds develop from ovules after double fertilization, involving the fusion of two pairs of gametes: one haploid sperm fuses with the haploid egg cell to form the diploid embryo whereas the second haploid sperm unites with the homo-diploid central cell and develops into the triploid endosperm. The endosperm is a placenta-like tissue supporting the development of the embryo by providing nutrients. Both fertilization products are surrounded by the protective seed coat, a maternal tissue derived from the integuments of the ovule. In the 1960s, Jerry Kermicle studied various alleles of the *colored1* (*r1*) locus in maize. The *R-r:standard* allele (hereafter referred to as *R*) controls the production of anthocyanin pigments in the aleurone, the outermost layer of the endosperm in maize seeds (kernels). If crossed to the recessive null allele *r-g* (hereafter referred to as *r*), Kermicle obtained fully colored kernels if *R* was inherited from the mother and mottled kernels with purple-brownish spots when *R* was inherited from the father (◘ Fig. 5.1).

Such a difference in phenotype depending on the direction of the cross indicates a parental effect. In this case, there is a maternal effect on *R* expression as most aleurone cells do not express *R* if it is inherited from the father. However, there are several fundamentally different mechanisms that could lead to such a maternal effect and Kermicle used an elegant series of genetic experiments to distinguish between them (Kermicle 1970).

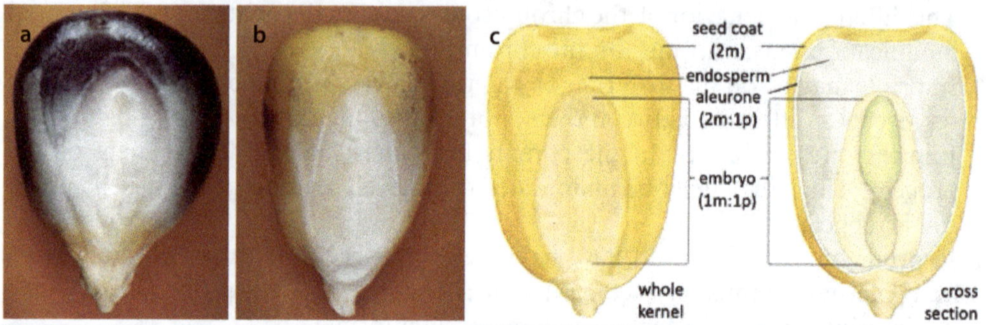

d Female	Male	Endosperm genotype (m:p)	Kernel phenotype
R R	*r r*	*R R:r*	
r r	*R R*	*r r:R*	
r r	*R* TR	*r r:*TR TR	
		r r:R	
R TR	*r r*	TR TR:*r*	
		R R:r	
R R, *mdr1 mdr1*	*r r*, *mdr1 mdr1*	*R R*, *mdr1 mdr1:r*, *mdr1*	
R R, *mdr1 mdr1*	*r r*, *MDR 1 MDR1*	*R R*, *mdr1 mdr1:r*, *MDRr1*	

Fig. 5.1 Imprinting at the *r1* locus. **a** Kernel with maternal (full pigmentation) and **b** with paternal inheritance of the *R* allele (mottled pigmentation) after a cross with the recessive, colorless *r* allele. (From Evans and Grossniklaus 2008). **c** Schematic representation of a maize kernel showing the composite structure of the seed with the purely maternal (2m) seed coat, the triploid endosperm including the pigmented aleurone with 2 maternal and 1 paternal genomes (2m:1p), and the diploid embryo (1m:1p). (From Encyclopedia Brittanica). **d** Results of crosses involving *R* and *r* alleles as well as the T*R* translocation where the long arm of A-chromosome 10 containing the *R* locus is translocated onto a non-essential B-chromosome. Such B-A translocation stocks can produce sperm cells with either two T*R*s or none, allowing the generation of kernels with different *R* dosage. The *maternal derepression of r1* (*mdr1*) mutation disrupts a *trans*-acting factor of imprinting required for full expression of the maternally inherited *R* allele. (From Baroux et al. 2002)

The maternal effect could be due to a dosage effect in the triploid endosperm as it contains two chromosome sets from the mother and only one from the father. Therefore, the copy number of *R* differs in reciprocal crosses and it is possible that full pigmentation requires a threshold that is only reached if two *R* copies are present. To test this hypothesis, Kermicle used B-A translocations (► Box 5.1) to generate kernels with two paternal copies of *R*. As the phenotype remained mottled, he excluded that different dosage was the basis of the maternal effect (◘ Fig. 5.1).

Methods Box 5.1: Genetic Analyses Using Translocation Stocks (◘ Box Fig. 5.1)

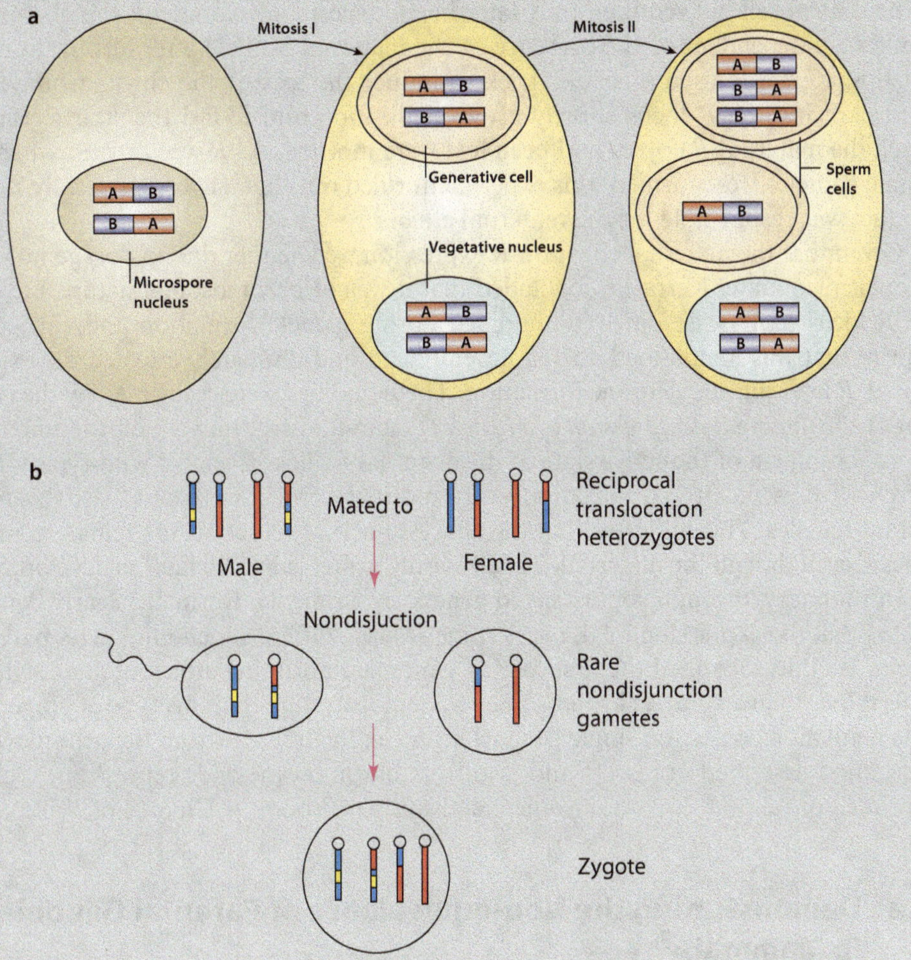

◘ Box Fig. 5.1 In both maize and mice, genetic studies that were crucial for the demonstration of genomic imprinting relied on translocation stocks. **a** Maize B-chromosomes do not carry any essential genes but can participate in translocations with the normal A-chromosomes. In the meiotic product (microspore) shown below, only the reciprocal A-B and B-A translocation chromosomes are shown. Microspores undergo two mitotic divisions to form the sperm. In mitosis I, the vegetative and generative lineage separate, while in mitosis II, the generative cell divides once more to form the two sperm cells that will participate in double fertilization. The B-chromosome can impart a peculiar behavior onto the translocated part of the A-chromosome that does not contain a centromere: during mitosis II, non-disjunction produces one sperm with two and one without a B-A translocation chromosome, allowing the manipulation of dosage for the genes present on the translocated part of the A-chromosome. (From Walbot and Evans 2003). **b** In mice, balanced reciprocal translocation heterozygotes were used to map chromosomal regions that contain imprinted loci. Crossing such translocation stocks carrying appropriate markers (yellow segment in the example below) generates some progeny that have a fully balanced genome, i.e., all chromosomal regions are present in two copies, but the translocated segments (distal parts of red and blue chromosomes) are of the same parental origin. If such progeny shows phenotypic abnormalities, the corresponding segments must contain imprinted genes required for normal development or behavior. (From Oakey and Beechy 2002)

Alternatively, the maternal effect could be of cytoplasmic nature, i.e., it depends on organelles, proteins, or RNAs produced prior to fertilization and deposited in the female gametes. If full pigmentation depended on a product that is maternally stored in the central cell but required only later in endosperm development, the phenotype would be independent of the presence of the maternal *R* allele after fertilization. To test this, Kermicle used a genetic trick to induce the loss of the chromosome fragment carrying the *R* locus shortly after fertilization and found that the sectors in which the maternal *R* copies had been lost were mottled. As *R* was present when the female gametes were formed, this experiment ruled out that a cytoplasmically stored product was responsible for the maternal effect.

Given that the maternal effect on *R* expression was neither due to dosage nor of a cytoplasmic nature, Kermicle concluded that it was of chromosomal nature, i.e., that the *R* locus itself was somehow modified during gamete formation and carried an imprint that was maintained during development and ultimately affected the expression of *R* later during aleurone formation. These findings were confirmed by the isolation of the *maternal derepression of r1* (*mdr1*) mutant, disrupting a factor required for the establishment of the active state of the maternal *R* allele. Even if a wild-type *MDR1* copy was present after fertilization, only mottled kernels were produced by *mdr1* mutant females. This indicates that the active state of *R* is set during female gametogenesis, even though the maternal *R* allele is only expressed much later in development.

In summary, through sophisticated genetic experiments, Kermicle clearly demonstrated that the expression of the very same *R* allele differs depending on its parental origin. He thus identified the first case of genomic imprinting affecting an individual locus rather than an entire chromosome set, demonstrating that this type of epigenetic gene regulation can act on single genes. This is clearly different from the genome-wide effects first described in insects and also from chromosome-wide repression, such as imprinted or random X chromosome inactivation (see book ▶ Chap. 4 of Wutz).

5.1.3 Demonstrating the Non-equivalence of Parental Genomes in Mammals

In mammals, chromosome-wide imprinting was also described in the 1970s when X-chromosome inactivation (see book ▶ Chap. 4 of Wutz) was found to be paternal-specific in marsupials and in the extraembryonic tissues of the mouse (Cooper et al. 1971). Furthermore, using translocation stocks (▶ Box 5.1), mouse geneticists identified regions with parental-specific phenotypes and described a maternal effect in the "hairpin-tail" mouse. However, they did not conclude that this was due to genomic imprinting but favored the interpretation of a cytoplasmic maternal effect. This view only changed with the seminal nuclear transfer experiments performed in Davor Solter's and Azim Surani's groups, which unequivocally showed that maternal and paternal genomes of the mouse are not equivalent (McGrath and Solter 1984; Surani et al. 1984).

After fertilization, the nucleus contributed by the sperm decondenses and forms a pronucleus that is larger than the female pronucleus, making them easily distinguishable. As these pronuclei remain observable for more than 12 hours, the parental genomes in the zygote can be manipulated. Using micropipettes to remove and add pronuclei, Surani and Solter generated zygotes that had either two maternal, two paternal, or one maternal and paternal pronucleus each (◘ Fig. 5.2).

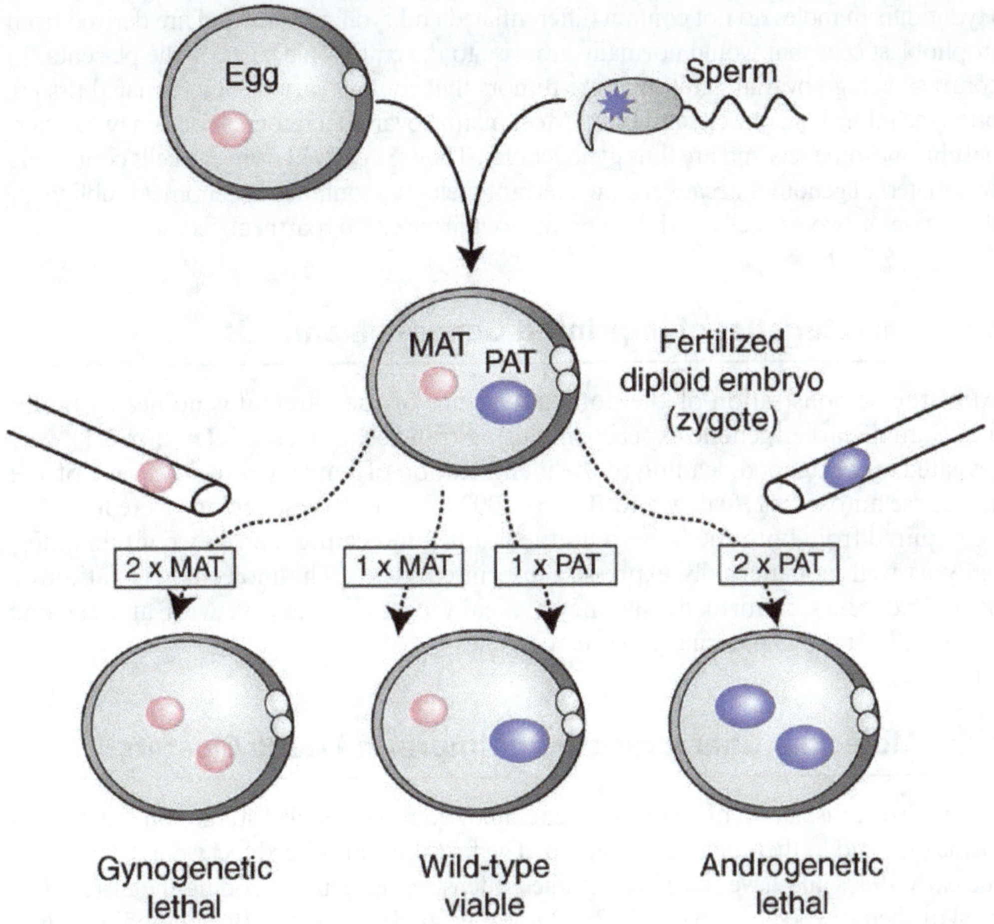

Fig. 5.2 Both maternal and paternal genomes are required for normal embryogenesis in the mouse. The zygote contains a maternal (pink) and paternal (purple) pronucleus that can be manipulated using a micropipette. By removing and adding different pronuclei, the following combinations were generated: gynogenetic and androgenetic embryos containing two maternal or two paternal pronuclei, respectively, aborted during embryogenesis, while embryos, in which the wild-type situation with one paternal and maternal pronucleus was recreated, developed normally. (From Barlow and Bartolomei 2014)

Only embryos with a biparental constitution survived and produced viable pubs, while gynogenotic (two maternal genomes) and androgenotic (two paternal genomes) embryos died early during embryogenesis. These results clearly demonstrate that both, maternal and paternal genomes are required for normal embryogenesis and suggest that the two parental genomes are not equivalent. Unlike in genetic crosses, neither genome dosage nor the cytoplasm of the zygote differed in these experiments, pointing to the existence of imprinted genes with only the paternally or maternally inherited allele being expressed. Nevertheless, it was only with the cloning of three imprinted genes and the demonstration of their parental-specific expression in the early 1990s that the idea of genomic imprinting in mammals gained widespread acceptance.

Conceptuses that develop from zygotes with only paternal or maternal genomes also spontaneously occur in humans and lead to abnormal pregnancies. Hydatidiform moles are derived from androgenetic zygotes that were generated by fertilization of an anuclear egg cell either by one sperm followed by chromosome doubling or by two sperm cells.

Hydatidiform moles do not contain differentiated embryonic tissues and are derived from trophoblast cells that would normally give rise to the embryonic parts of the placenta. In contrast, benign ovarian teratomas are tumors that contain various differentiated tissues, often including hair, muscle, and bone. Most mature ovarian teratomas contain two maternal chromosome sets and are thus gynogenotes. They are derived from egg cells containing two maternal genomes because of an aberrant meiosis, spontaneous genome doubling, or the fusion of two egg cells, and develop without fertilization (parthenogenesis).

5.2 Characteristics of Imprinted Genes in Mammals

After the demonstration of the non-equivalence of the parental genomes by Solter and Surani, earlier genetic experiments using translocation stocks (▶ Box 5.1) were revisited and extended, leading to the identification of imprinted regions on 8 of the 19 mouse autosomes (Oakey and Beechy 2002). Most of these regions were found to be required from both the father and the mother, indicating that they contain paternally as well as maternally expressed imprinted genes. The molecular isolation of imprinted genes confirmed that they typically occur in clusters with at least one maternally and one paternally expressed gene.

5.2.1 Molecular Characteristics of Imprinted Gene Clusters

The molecular isolation of imprinted genes allowed more detailed studies on the mechanisms that lead to their unique regulation. The first three imprinted loci were described in the early 1990s and have since been studied intensely: the genes encoding the insulin-like growth factor type-2 receptor (*Igf2r*, Barlow et al. 1991), the *H19* non-coding RNA (Bartolomei et al. 1991), and the insulin-like growth factor 2 (*Igf2*; DeChiara et al. 1991). To date, about 150 genes are known to be imprinted in the mouse. Further characterization has shown that not all imprinted genes are exclusively expressed from only one parental allele; rather, many show a consistent, biased expression of one parental allele. Thus, the definition of genomic imprinting should be broadened to include all genes that show non-equivalent expression from the parental alleles. Although many imprinted genes are expressed in the placenta, some show rather specific expression patterns, often restricted to adult tissues such as the brain. This emphasizes a key point initially made by Metz: while the imprints are established during gametogenesis, their readout, i.e., the biased or specific expression of one of the parental alleles, can manifest itself much later in development. Thus, the imprint itself and its effect on gene expression should not be confounded. Some genes show imprinted expression in some tissues but are biallelically expressed in others. Imprinted expression can also be stage-specific, whereby even the parental association of the active and silenced alleles can be reversed in different tissues. In other words, the effect of the imprint that has been established in the germline differs depending on developmental stage and tissue. These observations suggest complex mechanisms that translate the imprinting mark into a discernable effect on gene expression and ultimately the phenotype.

A typical imprinted gene cluster contains both maternally and paternally expressed genes and usually also biparentally expressed ones (◘ Fig. 5.3). The expression of the imprinted genes is controlled by an imprinting control element, also known as ICE. In most imprinted gene clusters, the ICE shows differential DNA methylation,

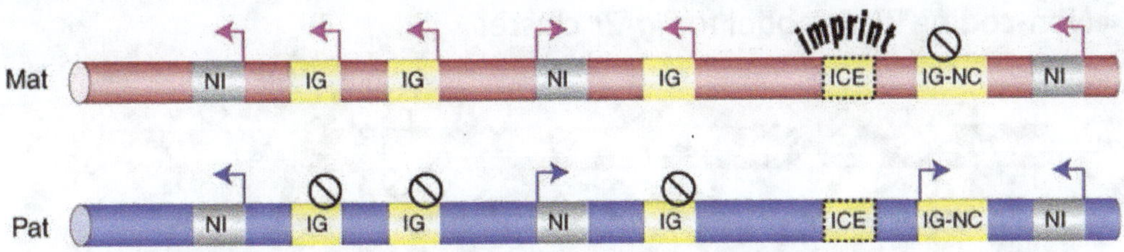

Fig. 5.3 A typical imprinted gene cluster with non-imprinted genes (NI) as well as paternally and maternally expressed imprinted genes (IG). Maternal (Mat) and paternal (Pat) chromosomes of a typical imprinted gene cluster are shown with silenced (slashed circles) and expressed (arrows) alleles. Most clusters contain at least one imprinted, non-coding RNA gene (IG-NC). Imprinted expression is controlled by an imprinting control element (ICE) that carries an epigenetic imprinting mark that was established in and inherited from one parental gamete. The ICE often controls expression of the imprinted IG-NC, which plays an important role in regulating the cluster. (From Barlow and Bartolomei 2014)

which was established in only one of the parents during gametogenesis (gametic or primary imprint) and then maintained exclusively on this allele throughout post-fertilization development. Thus, the gametic imprint marks the maternal or paternal alleles differently in the respective zygote. Initially, all studied imprinted gene clusters in the mouse carried a gametic imprint based on 5-methyl-cytosine (see book ▶ Chap. 1 of Wutz), but this is not true in macaques where several ICEs do not show differential DNA methylation in the gametes. More recently, it was shown that primary imprints can also be based on the histone modification H3K27me3 (Inoue et al. 2017). Deletion of the ICE leads to parent-of-origin-specific effects. Such a deletion has usually no effect on the expression of the genes on the allele carrying the imprinting mark. In contrast, deletion of the ICE on the allele without the gametic imprinting mark leads to a switch, such that the expression of the imprinted genes in the cluster now resembles that of the allele that is normally carrying the imprinting mark.

5.2.2 Molecular Mechanisms Leading to Imprinted Expression

The majority of imprinted gene clusters contains at least one imprinted gene for a non-coding RNA (IG-NC). While several imprinted non-coding RNA genes are very long, ranging from about 100 kb for *Airn* to over 1000 kb for *LNCAT*, the *H19* non-coding RNA is only 2.3 kb. Two observations suggest that the IG-NC plays a role in regulating imprinted expression. First, the IG-NC usually shows the opposite parental-specific expression of the protein-coding genes in the cluster, and second, the ICE of the cluster carrying the gametic imprint typically overlaps with the IG-NC promoter, controlling its expression (Fig. 5.4). The role of the long non-coding RNA in regulating imprinted expression was tested in several clusters by truncating it through the insertion of a polyadenylation signal at the endogenous locus. Truncation of *Airn*, *Kcqnt1ot1*, and *Nespas* showed that expression of these long non-coding RNAs was necessary to silence the paternal alleles of the protein-coding genes in the cluster. How exactly they silence other genes in *cis* is currently not known, although models involving RNA-interference (see book ▶ Chap. 6 of Grossniklaus), transcriptional interference, or coating of the local chromatin and recruitment of repressive marks similar to X-chromosome inactivation (see book ▶ Chap. 4 of Wutz) have been proposed. In all three clusters, the maternal allele carries the gametic DNA methylation imprint that prevents transcription of the IG-NC (Fig. 5.4a).

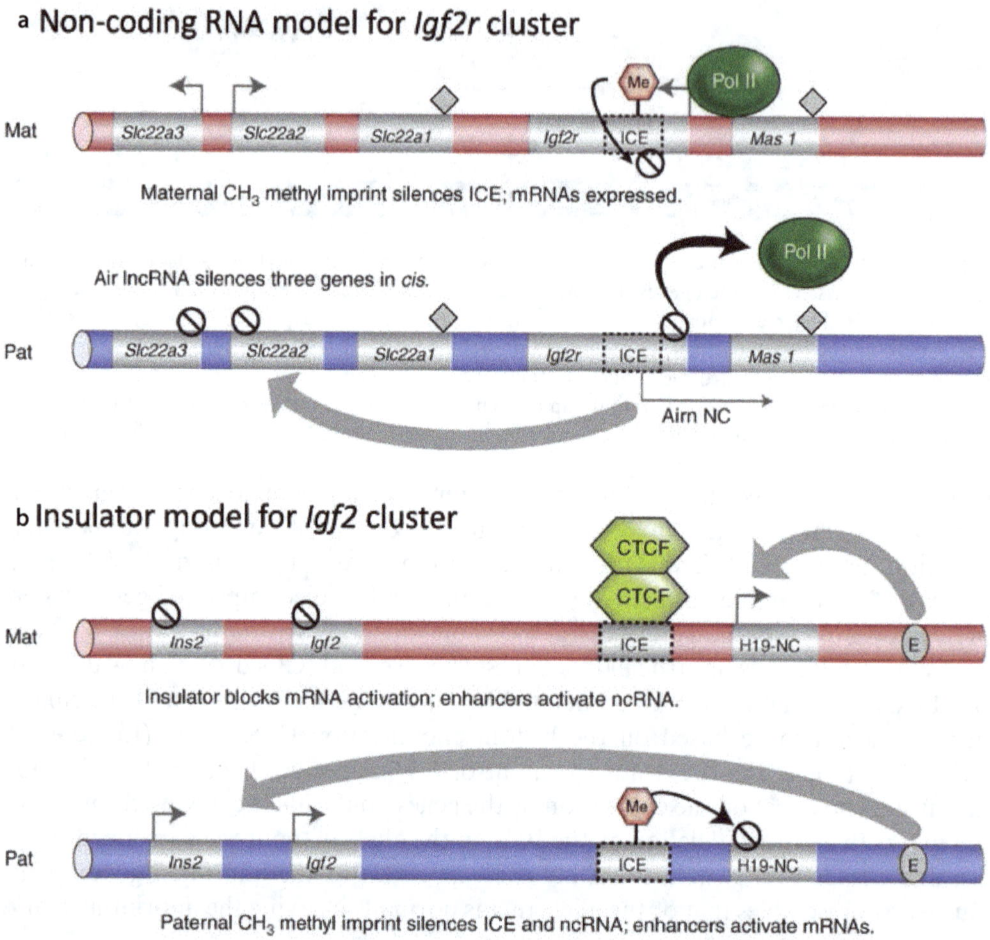

Fig. 5.4 Molecular mechanisms underlying parental-specific expression in imprinted gene clusters. **a** In the *Igf2r* cluster, the ICE carrying a DNA methylation imprint overlaps with the promoter of the *Airn* non-coding RNA (Airn NC). Expression in the placenta is shown where *Slc22a1* and *Mas1* are not expressed (diamonds). On the maternal (Mat) chromosome, the *Airn* promoter is methylated and silenced (slashed circle), while *Igf2r*, *Slc22a2*, and *Slc22a3* are expressed (arrows). On the paternal (Pat) chromosome, lacking the gametic methylation imprint, *Airn* is expressed and silences *Igf2r*, *Slc22a2*, and *Slc22a3* in *cis*. **b** The genes of the *Igf2* cluster are regulated by the same enhancers (E) distal to the *H19* non-coding RNA gene (H19-NC). On the maternal chromosome, the ICE is unmethylated, allowing CTCF proteins to bind and form an insulator that blocks the enhancers from interacting with the promoters of *Igf2* and *Ins2*, while they can interact and activate *H19*. On the paternal chromosome, the ICE carries a gametic methylation imprint which prevents the binding of CTCF and, thus, the formation of an insulator. The enhancers now preferentially interact with and activate the *Igf2* and *Ins2* promoters, while *H19* is not expressed. (From Barlow and Bartolomei 2014)

In contrast, at the *Igf2* cluster the gametic methylation imprint is on the paternal allele and deletion of the *H19* non-coding RNA gene had no effect on silencing the maternal alleles of *Igf2* and *Ins2*. The *Igf2* cluster uses an insulator-based mechanism that results in parent-of-origin-dependent expression (Fig. 5.4b). All genes in the cluster are regulated by the same enhancers located downstream of the *H19* gene. As described in ► Chap. 1, chromatin structure and topology contribute to the regulation of gene expression (see book ► Chap. 1 of Wutz). In particular, CTCF was implicated in the formation of insulators and chromatin loops. The unmethylated

ICE, located upstream the *H19* gene, serves as a binding site for CTCF proteins, which form an insulator that blocks the interaction of these enhancers with the *Igf2* and *Ins2* promoters, while activating the *H19* promoter. If the ICE is methylated, CTCF proteins cannot bind and the insulator does not form, thus allowing the interaction of the enhancers with the more distant promoters of the protein-coding genes *Igf2* and *Ins2*. CTCF binding sites were also identified at other imprinting clusters, where a similar mechanism may regulate imprinted expression.

5.2.3 The Life Cycle of a Genomic Imprint

As described above, essentially all imprinted gene clusters carry differentially methylated regions (DMRs) as well as regions that differ with respect to H3K27me3 on the maternal and paternal chromosomes. However, not all of these are necessarily the primary or gametic imprints that mark the parental chromosomes and are typically located at ICEs. As DNA methylation is associated with repression, DMRs can also simply coincide with the silenced alleles in a cluster, as can other repressive marks such as H3K27me3. To identify the primary imprint, whether it is differential DNA methylation or a chromatin modification, it is therefore important to determine when the imprinting mark is established during development. If it forms during gametogenesis and is maintained throughout development, it is likely the primary imprinting mark that confers a memory of its parent-of-origin. On the other hand, if the mark is established after fertilization, when the two parental genomes are in the same nucleus, it might be considered a consequence of the silenced state of imprinted genes.

As imprints have to be reset in each generation, they undergo a life cycle of erasure and reestablishment in the germline. This occurs in primordial germ cells where DNA methylation is erased genome-wide and then reestablished, with DMRs getting methylated according to the sex of the individual (◘ Fig. 5.5a). There are many more DMRs that are methylated on the maternal allele (22 of 25 well-studied DMRs) than on the paternal one, and the timing of methylation differs between males and females. Maternal methylation imprints are established during oocyte maturation in meiotic prophase, which occurs after birth. Paternal methylation imprints are established in the developing testes of the fetus in prospermatogonia long before the germ cells enter into meiosis. Despite this difference in timing, the acquisition of gametic methylation imprints depends on the *de novo* DNA methyltransferase 3A (Dnmt3A, see book ► Chap. 1 of Wutz). In oocytes, Dnmt3A is aided by the non-catalytic Dnmt3L that directs DNA methylation to transcriptionally active regions, which are enriched in H3K36me3 but devoid of H3K4me2/3 (Stewart et al. 2016). In the male germline, *de novo* DNA methylation is linked to Piwi-interacting RNAS (piRNAs) that direct it to repeats and transposable elements (see book ► Chap. 6 of Grossniklaus). This targeting mechanism includes the paternal gametic imprint in the *Rasgrf1* cluster, which contains a retrotransposon. In addition to Dnmt3A/L, Dnmt3B is required for methylation at the *Rasgrf1* DMR but it plays no role for the maternal methylome. In sperm, DNA methylation is more evenly distributed in both genic and intergenic regions. Why Dnmt3A/L appears to act rather unspecifically in sperm but is precisely targeted to transcribed genes in oocytes is currently not understood.

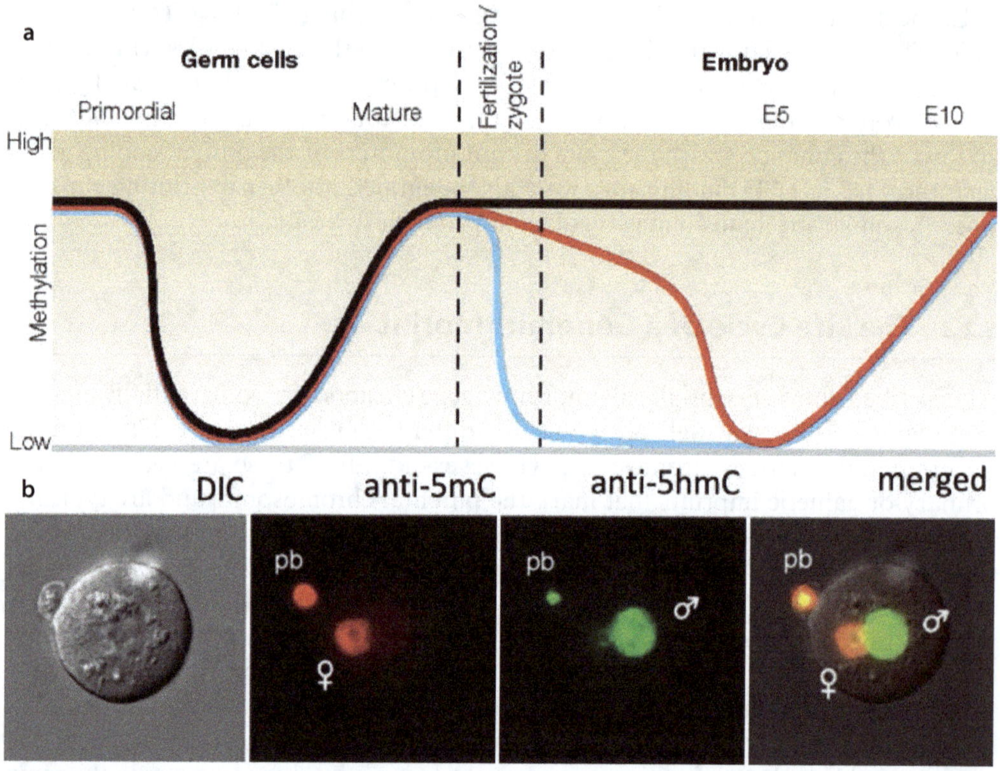

Fig. 5.5 Reprogramming of DNA methylation in the primordial germ cells and the embryo. **a** The schematic shows how the level of methylation in non-methylated (grey) and methylated (black) DMRs of imprinted genes and in non-imprinted sequences (red: maternal, blue: paternal genome) changes during development. The x-axis indicating developmental timing and the y-axis showing relative methylation levels are not to scale. In germ cells, the methylation imprints are erased and reset according the sex of the individual. In the zygote and early embryo, there is global demethylation that, however, does not affect the gametic imprints, which are maintained. The paternal genome loses DNA methylation in the zygote very rapidly, even before the first division. In contrast, the maternal genome loses DNA methylation gradually, possibly through passive demethylation by DNA replication. (From Reik and Walter 2001). **b** The rapid loss of 5-methyl-cytosine (5mC) in the male pronucleus occurs concomitantly with oxidation, resulting in 5-hydroxymethyl-cytosine (5hmC). As revealed by antibody staining, the female pronucleus und polar body (pb) still show high levels of 5mC when the male pronucleus has completely lost this antigen but gained 5hmC. (From Iqbal et al. 2011)

Once the gametic imprints are established, they are propagated over cell divisions by the maintenance DNA methyltransferase Dnmt1, which methylates hemimethylated CpG sites directly at the replication fork (see book ► Chap. 1 of Wutz). However, imprinted loci are challenged by a second genome-wide reprogramming event immediately following fertilization, involving demethylation during preimplantation development and *de novo* methylation in the postimplantation embryo (Fig. 5.5a). This wave of genome reprogramming is postulated to allow the resetting of the zygotic genome to a totipotent state. The parental genomes that contribute to the zygote are derived from highly differentiated gametes and, therefore, carry epigenetic markings that have to be removed in order to generate totipotent cells and, subsequently, all cell types of the organism (see also book ► Chap. 7 of Paro).

While the paternal genome undergoes a rapid loss of DNA methylation even before the first division, the maternal genome is demethylated more slowly during preimplantation development (◘ Fig. 5.5a). These kinetics indicate an active demethylation of the paternal genome whereas the maternal genome likely loses DNA methylation passively in a DNA replication-dependent manner (see book ► Chap. 1 of Wutz). Active demethylation was proposed to be mediated by ten-eleven translocation methylcytosine dioxygenases (TETs), which can oxidize 5-methyl-cytosine (5mC) to 5-hydroxymethyl-cytosine (5hmC) and further to 5-formyl- and 5-carboxyl-cytosine (see book ► Chap. 1 of Wutz). Indeed, TET3 expression coincides with loss of 5mC and the accumulation of 5hmC in the male pronucleus (◘ Fig. 5.5b), leading to the widely held view that demethylation occurs via oxidation by TET enzymes. However, recent studies have shed some doubt on this notion as demethylation seemed complete prior to the accumulation of 5hmC and still occurred in an oocyte-specific *Tet3* knock-out mouse (SanMiguel and Bartolomei 2018).

One of the most intriguing questions is how gametic imprints escape demethylation in the embryo and unmethylated DMRs are protected from becoming methylated when the rest of the genome regains DNA methylation. This is probably achieved by *trans*-acting factors that bind to specific *cis*-regulatory regions to protect gametic DMRs. One of these proteins is DPPA3, which is responsible for the maintenance of DNA methylation at some of the gametic DMRs. Another protein that protects methylation imprints is ZFP57, a zinc finger protein with binding preference for TGCCGC, which occurs at most gametic DMRs involved in imprinting. ZFP57 binds only to methylated sites and interacts with the KAP1 protein that recruits various repressive histone modifying enzymes and, importantly, Dnmt1, which maintains methylation at the DMR. In maternal-zygotic *Zfp57* mutant embryos, there is a loss of DNA methylation at many gametic DMRs accompanied by extensive changes in the expression of imprinted genes, thus illustrating its importance in protecting the DMRs during reprogramming (SanMiguel and Bartolomei 2018).

5.3 Genomic Imprinting and Human Disease

Imprinted genes are expressed in diverse tissues and their functions are manifold. Consistent with the expression of many imprinted genes in the embryo and placenta, mutations often affect fetal growth and development. Other imprinted genes show preferential expression in the brain and mutations in the mouse can lead to the development of behavioral phenotypes. Given that only one of the parental alleles of imprinted genes is active, it is not surprising that mutations in imprinted gene clusters are often linked to human disease (see book ► Chap. 8 of Santoro). While some of these diseases are clearly linked to a specific imprinting cluster, others show disruptions of imprinting at multiple loci across the genome. A selection of imprinting disorders is shown in ◘ Table 5.1 (based on Monk et al. 2019):

As becomes obvious from the selection in ◘ Table 5.1, many imprinting disorders affect the same genomic region but lead to distinct clinical features depending on which of the parental alleles is affected. For instance, in some patients with Silver-Russel Syndrome, the paternal copy of the *Igf2/H19* imprinting cluster is missing or affected, while in Beckwith-Wiedemann Syndrome, it is the maternal copy. Likewise, in Temple Syndrome, the paternal allele of the *Meg3-Dlk1* cluster is affected, while the maternal one is missing in most patients with Kagami-Ogata Syndrome.

Table 5.1 Selected human disorders related to defects in imprinting clusters

Imprinting disorder	Chromosome	Affected imprinting cluster
Silver-Russel Syndrome	11p15.5 7p11.2 Others	*Kcnq1, Igf2/H19* *Grb10* ...
Beckwith-Wiedemann Syndrome	11p15.5	*Kcnq1, Igf2/H19*
Temple Syndrome	14q32	*Dlk1-Dlo3*
Kagami-Ogata Syndrome	14q32	*Dlk1-Dlo3*
Prader-Willi Syndrome	15q11-q13	*Snrpn/Ube3A*
Angelman Syndrome	15q11-q13	*Snrpn/Ube3A*

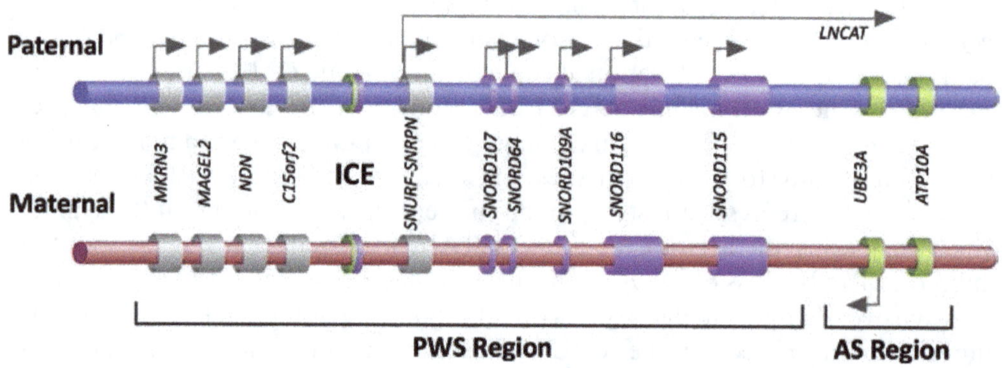

Fig. 5.6 The *Snrpn/Ube3A* imprinted gene cluster associated with Prader-Willi Syndrome (PWS) and Angelman Syndrome (AS). The ICE controls imprinted expression of the genes in the cluster. The *SNURF-SNRPN* gene encodes not only two protein products, Snurf and Snrpn, but also a paternally expressed long non-coding antisense transcript (*LNCAT*) that overlaps with the promoter of the maternally expressed *UBE3A* gene responsible for AS. (From Zoghbi and Beaudet 2016)

To illustrate the molecular basis of imprinting disorders, we focus on the Prader-Willi Syndrome (PWS) and Angelman Syndrome (AS), both affecting the *Snrpn/Ube3A* imprinted gene cluster. Among other clinical features, PWS is characterized by a lower birth weight, neonatal hypotonia, postnatal growth restriction followed by hyperphagia and obesity, short stature, hypogonadism, and mild intellectual disability. In contrast, AS patients can have an above-average birth weight (depending on the exact lesion), severe delay in postnatal development, ataxia, scoliosis, microcephaly, seizures, severe intellectual disability, minimal verbal skills, and a happy affect with frequent bouts of laughter.

The majority of patients with either disorder carry a deletion of about 5–6 Mb in 15q11. While the deletion is inherited paternally in PWS patients, it is derived from the mother in AS children. This difference explains the distinct clinical features in PWS and AS patients, as only the maternal or paternal alleles of the *Snrpn/Ube3A* cluster, respectively, are expressed. Apart from deletions, other genetic lesions leading to loss-of-function mutations of genes in the *Snrpn/Ube3A* cluster can cause the syndromes (Fig. 5.6). For AS, about 10–15% of the cases carry a mutation in the maternally expressed *UBE3A* gene, identifying *UBE3A* as gene likely responsible for AS. It is not yet fully clear which of the genes in the PWS region of the imprinting cluster is responsible for PWS, although recent evidence points to the *SNORD116* locus encoding a set of small nucleolar RNAs.

Apart from genetic causes of these syndromes, there are also cases where the *Snrpn*/*Ube3A* cluster is intact but its genes are epigenetically misregulated. This is for instance the case if both homologs of chromosome 15 originate from the same parent, a condition called uniparental disomy (UPD). Maternal UPD is responsible for PWS, whereas AS is caused by paternal UPD. The origin of UPDs lies in non-disjunction during female meiosis that, after fertilization, leads to zygotes that either have only one paternal (monosomy) or two maternal and one paternal (trisomy) chromosome 15 in the case of PWS and AS. These aberrant zygotes can develop normally if they spontaneously loose or duplicate chromosome 15. Loss of the paternal copy from trisomic zygotes leads to maternal UPD, while duplication of the single paternal chromosome in monosomic zygotes to paternal UPD. As in individuals with UPD the two homologs stem from the same parent, they lack expression of imprinted genes specific to the other parent, resulting in imprinting disorders.

Importantly, PWS and AS can also stem from imprinting defects. Imprinted expression at the *Snrpn*/*Ube3A* cluster is controlled by an ICE upstream of the *SNURF-SNRPN* gene. If a mutation affects the function of the ICE, the genes in the cluster lose their parent-of-origin-specific expression. The same can result from a failure to reprogram the gametic imprint in primordial germ cells. For instance, if the methylation mark on the maternal ICE is not erased in prospermatogonia, a male will transmit it with a maternal rather than a paternal gametic mark. This will lead to the formation of a zygote with two "maternal" *Snrpn*/*Ube3A* clusters causing PWS. Although such imprinting defects are responsible for only a minority of the PWS and AS cases, they contributed important insights into the resetting of imprints and the regulation of imprinted expression at the *Snrpn*/*Ube3A* cluster.

5.4 Genomic Imprinting in Flowering Plants

Although imprinting of individual genes was originally discovered in plants and the parent-of-origin-dependent expression of a few genes was documented in the maize endosperm, the mechanisms underlying genomic imprinting were not well studied until the discovery of imprinted genes in the model plant *Arabidopsis thaliana* in the late 1990s. Plant imprinting studies were reinvigorated by the discovery of *MEDEA* (*MEA*) (Grossniklaus et al. 1998), a maternally expressed imprinted gene encoding a homolog of *Drosophila melanogaster Enhancer of zeste*, the histone methyltransferase in the conserved *Polycomb* Repressive Complex 2 (PRC2; see book ▶ Chap. 3 of Paro). The *mea* mutant was identified in a screen for maternal effect mutants, with the genotype of the female gametes determining the phenotype of the developing seeds. Because the gametes of plants are produced by haploid multicellular organisms (gametophytes) that form through mitotic divisions after meiosis, such mutants show a peculiar genetic behavior. In a heterozygous *mea*/*MEA* plant, half the meiotic products will carry the mutant *mea* allele and half the wild-type *MEA* allele. Correspondingly, also half of the gametophytes and the gametes they contain (egg and central cell in the female gametophyte; two sperm cells in pollen, the male game-

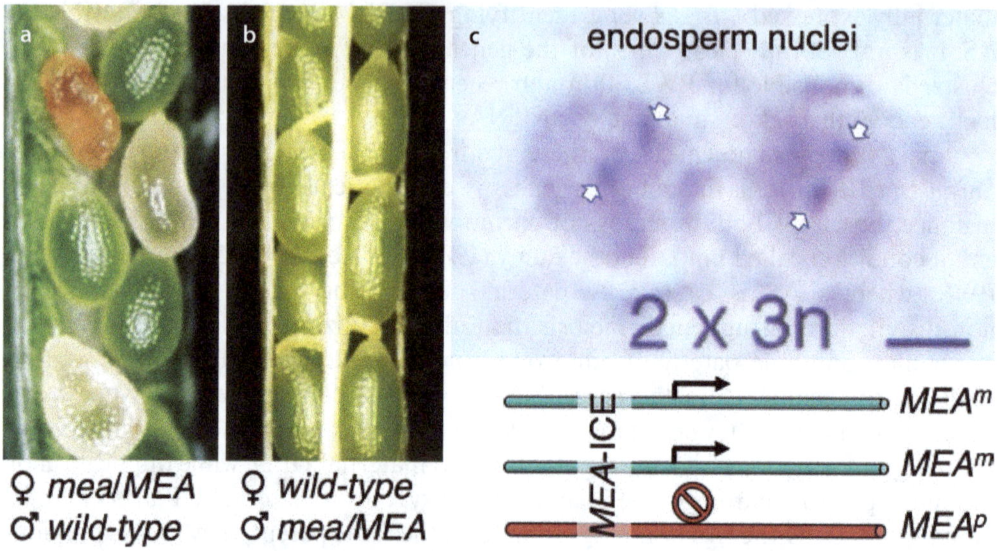

Fig. 5.7 Genomic imprinting is responsible for maternal effect seed abortion observed in the *mea* mutant. **a** and **b** show opened siliques with developing seeds from reciprocal crosses. If the mutant *mea* allele is inherited from the mother, seeds abort (red and brown seeds in **a** but if it is inherited from the father, it has no effect on seed development **b**. (From Grossniklaus et al. 1998). **c** Detection of nascent transcripts at the *MEA* locus (arrows) show that only two of the three alleles in the endosperm are actively transcribed, i.e., the two maternal copies as illustrated schematically. (From Vielle-Calzada et al. 1999)

tophyte) will carry the mutant allele. Thus, if a *mea/MEA* heterozygous female plant is crossed with a wild-type male, half of the seeds abort while, in the reciprocal cross, all seeds develop normally (Fig. 5.7).

Of course, similar to kernel pigmentation regulated by the *R* allele in maize, maternal effect seed abortion observed in *mea* mutants could be caused by (i) a dosage effect in the endosperm, (ii) a missing cytoplasmically stored product, or (iii) genomic imprinting. As *A. thaliana* does not offer the sophisticated genetic tools available in maize, a dosage effect was excluded by generating tetraploid plants that produce endosperm with *mea:MEA* allelic ratios ranging from 6:0 to 0:6. Distinction of the remaining two possibilities required the cloning of *MEA* and analysis of its allelic expression using parental-specific polymorphisms. This showed that only maternal transcripts could be detected in developing seeds. However, *MEA* was already expressed in the female gametes, such that these maternal transcripts could have either been produced exclusively before fertilization and stored in the egg cell or, in addition, also be derived from the maternal allele after fertilization. RNA *in situ* hybridization to detect nascent transcripts in triploid endosperm nuclei showed that only two of the three *MEA* alleles were transcribed. Nascent transcripts temporarily remain at their chromosomal site of transcription forming “nuclear dots” of high transcript concentration (Fig. 5.7). Thus, while *MEA* is already expressed in the female gametes, it is also regulated by genomic imprinting leading to the transcription of only the maternal *MEA* alleles in the endosperm (Vielle-Calzada et al. 1999). The discovery of genomic imprinting in *A. thaliana* stimulated many studies both at the genome-wide level as well as on the molecular mechanisms underlying parental-specific expression at specific loci in plants.

5.4.1 Genomic Imprinting Occurs Predominantly in the Endosperm But Also Exists in the Embryo

As all imprinted genes initially studied in maize were expressed in the endosperm (Messing and Grossniklaus 1999) and the DNA of the maternal genome was found to be hypomethylated in this tissue (Hsieh et al. 2009), imprinting is often referred to as endosperm-specific. Despite imprinted genes also having been identified in the plant embryo, most studies to date focused on endosperm. Over the last years, many genome-wide studies identified candidate imprinted genes in several species, including *A. thaliana*, maize, and rice (*Oryza sativa*). These studies relied on sequencing RNA isolated from dissected endosperm tissue or sorted endosperm nuclei, and subsequent searching for transcripts with parentally biased expression based on nucleotide polymorphisms between the parents.

All studies looked at steady-state levels of mRNA transcripts, which does not distinguish between mRNAs that were exclusively produced prior to fertilization and stored in the gametes and those that were differentially transcribed from the parental alleles after fertilization. Although endosperm was isolated several days after fertilization, the half-life of mRNAs is unknown and they could potentially be very stable. Thus, to conclude that a gene is regulated by genomic imprinting requires the demonstration of its active transcription after fertilization. Only in the zygote or fertilized central cell, both parental alleles will be present in the same nucleus and transcribed based on their gametic imprint. This would be the case if no transcripts were detected in the gametes, although cases like *MEA*, which is expressed in the female gametes as well as imprinted, would be missed by this strategy. Alternatively, active transcription can be shown by looking at nascent transcripts (◘ Fig. 5.7). While, in plants, this analysis has so far only been done for *MEA*, fluorescent *in situ* hybridization to detect RNAs (RNA-FISH) is a standard technique in mammals. In summary, while RNA sequencing experiments have identified hundreds of candidate imprinted genes in diverse species, regulation by genomic imprinting has unequivocally been demonstrated for only a few of them.

Moreover, the overlap of presumptive imprinted genes between genome-wide studies performed in different laboratories is very small (Wyder et al. 2019). It is possible that the hypomethylation of the maternal genome in the endosperm leads to stochastic changes in expression, which differ between individuals and are influenced by the environment, explaining the poor overlap between studies. This would mean that the parent-of-origin-dependent expression of these genes would be accidental and not play a crucial role in development. Consistent with this idea, the function of only a handful of imprinted genes is known, including *A. thaliana MEA* and *FERTILIZATION INDEPENDENT SEED2* (*FIS2*), which encodes a homolog of *Drosophila Su(z)12*. MEA and FIS2 are subunits of FIS-PRC2, a PRC2 variant (see book ► Chap. 3 by Paro) that is required for normal seed development.

Although most imprinting studies in plants focused on the endosperm, genomic imprinting has also been reported in the embryo. In *A. thaliana*, about a dozen candidate genes were identified, ten of which were not expressed in the gametes, indicating active expression after fertilization and thus *bona fide* imprinting. A recent study in maize identified several dozen candidate imprinted genes. As in mammals, gametic imprints of genes expressed in plant embryos would have to be reset in every generation. This is not true for imprints in the endosperm, which is a terminal tissue that does not contribute to the next generation. As in mammals, the primary imprints have to be established during gametogenesis when the parental genomes are separated. In both maize and *A. thaliana*, genes with parental allele-specific expression in young embryos become biallelically expressed later during embryogenesis. Whether this represents erasure of the gametic imprint or an alternative transcriptional regulation is currently unknown.

5.4.2 Mechanisms Underlying Imprinting Show Similarities Between Mammals and Plants

Both, the dissection of imprinting regulation at specific loci as well as genome-wide profiling studies of DNA methylation and histone modifications have shed light onto the mechanisms underlying genomic imprinting in plants. Again, studies on the regulation of the *MEA* locus paved the way to show that, as in mammals, DNA methylation plays a prominent role in genomic imprinting in plants. Unlike in mammals, where only CGs are methylated (see book ► Chap. 1 of Wutz), in plants, DNA methylation can occur in the CG, CHG, and CHH context (with $H \neq G$). While DOMAINS REARRANGED METHYLTRANSFERASE2 (DRM2) is the major *de novo* methyltransferase for all Cs, distinct DNA methyltransferases are responsible to maintain methylation in the different contexts: the Dnmt1 homolog MET1 for CG, the CHROMOMETHYLTRANSFERASE3 (CMT3) for CHG, and DRM2 together with CMT2 for CHH.

Mutants affecting *MET1* indicated an involvement of DNA methylation in silencing the paternal *MEA* allele. This was confirmed by the identification of DEMETER (DME), a DNA glycosylase that can excise methylated cytosines through base excision repair (Choi et al. 2002). DME is preferentially expressed in the central cell, where it leads to demethylation that then results in the hypomethylation of the maternal chromosomes in the endosperm after fertilization. In the *dme* mutant, the maternal *MEA* allele fails to be expressed, indicating that demethylation is required to set its active state. Regulation of imprinting by the maintenance methyltransferase MET1 and DME, which removes this epigenetic mark, has also been demonstrated for several other loci.

In addition to DNA methylation, H3K27me3 mediated by PRC2 was shown to regulate many imprinted genes. This was first shown for *PHERES1* (*PHE1*), a direct target of FIS-PRC2. The paternal allele of *PHE1* is expressed at a much higher level than the maternal one, which is specifically repressed by FIS-PRC2 during seed development (Köhler et al. 2005). The repression of the maternal *PHE1* allele also depends on a DMR at tandem repeats downstream of the gene. It has been proposed that the unmethylated DMR allows the formation of a repressive chromatin loop through interaction with FIS-PRC2 at the *PHE1* promoter (◘ Fig. 5.8). In its methylated form, the DMR prevents the binding of yet unknown factors involved in loop formation, resulting in an active paternal allele. While loops involved in imprinting await experimental confirmation in plants, the existence of such loops has been shown in mice. As discussed earlier, at the *Igf2*/*H19* imprinting cluster, CTCF binds to the unmethylated ICE and, together with PRC2, forms a chromatin

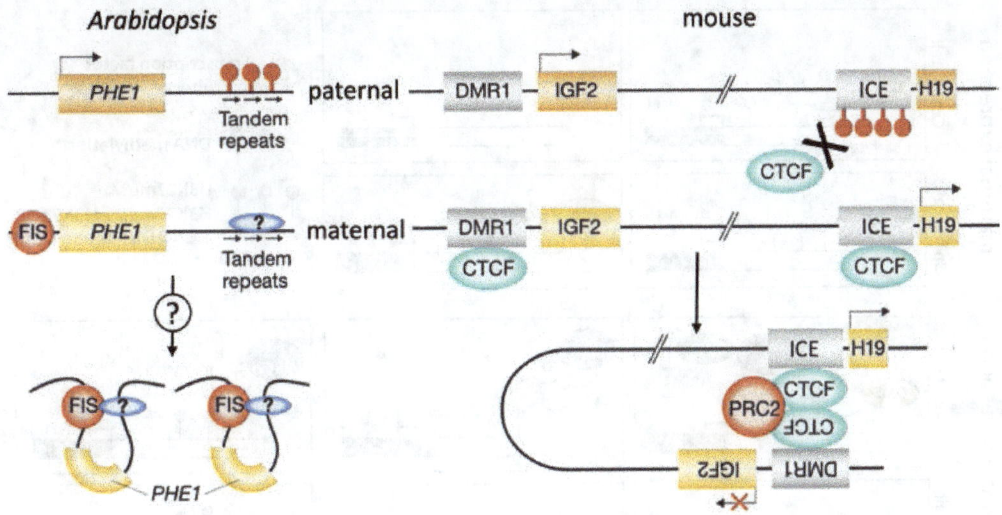

Fig. 5.8 Chromatin loops as part of the imprinting mechanism. On the left, the active paternal *PHE1* allele has methylated tandem repeats (DMR), while on the maternal allele, the DMR is unmethylated, which is proposed to allow the formation of a chromatin loop on the two maternal chromosomes in the endosperm. On the right, the *Igf2/H19* cluster is shown for which loop formation on the maternal chromosome, mediated by CTCF and PRC2, has been demonstrated. (From Loppin and Oakey 2009)

loop that represses *Igf2* (Fig. 5.8). PRC2 does not only play an essential role in seed development but also regulates some paternally repressed genes in the mouse placenta. Moreover, PRC2-mediated H3K27me3 can serve as gametic imprint, but relative to DNA methylation, the role of PRC2 is relatively minor (Barlow and Bartolomei 2014). Nevertheless, DNA methylation and PRC2 are involved in genomic imprinting in both mammals and plants, highlighting the independent recruitment of these two repressive modules in both lineages in the evolution of parental-specific expression.

Genome-wide profiling studies of DNA methylation, repressive chromatin marks, and gene expression in the endosperm have shown that a major determinant for maternally expressed imprinted plant genes is DNA methylation (Batista and Köhler 2020). In many cases, these genes are silenced by DNA methylation in most tissues and their demethylation by DME in the central cell leads to the expression of the maternal alleles in the endosperm. In other cases, *de novo* methylation of the paternal allele in sperm may lead to their imprinted maternal-specific expression (Fig. 5.9). For paternally expressed genes, histone modifications mediated by PRC2 (H3K27me3) or CHROMOMETHYLASE3 (CMT3) and plant SU(VAR)3-9 homologs (H3K9me2) play a major role in regulating imprinted expression. This can occur through demethylation of the maternal allele by DME followed by the deposition of repressive histone marks in combination with DNA methylation of the paternal allele, preventing the binding of PRC2 as it was shown for *PHE1* (Fig. 5.8). Alternatively, the maternal allele may be specifically targeted by recruiting PRC2 via selected transcription factors, or by removal of repressive histone marks of broadly silenced genes specifically in the sperm cell (Fig. 5.9).

While these mechanisms may explain the regulation of up to 30% and 90% of maternally and paternally expressed genes in the endosperm of *A. thaliana* (Batista and Köhler 2020), respectively, the regulation of many other imprinted genes does

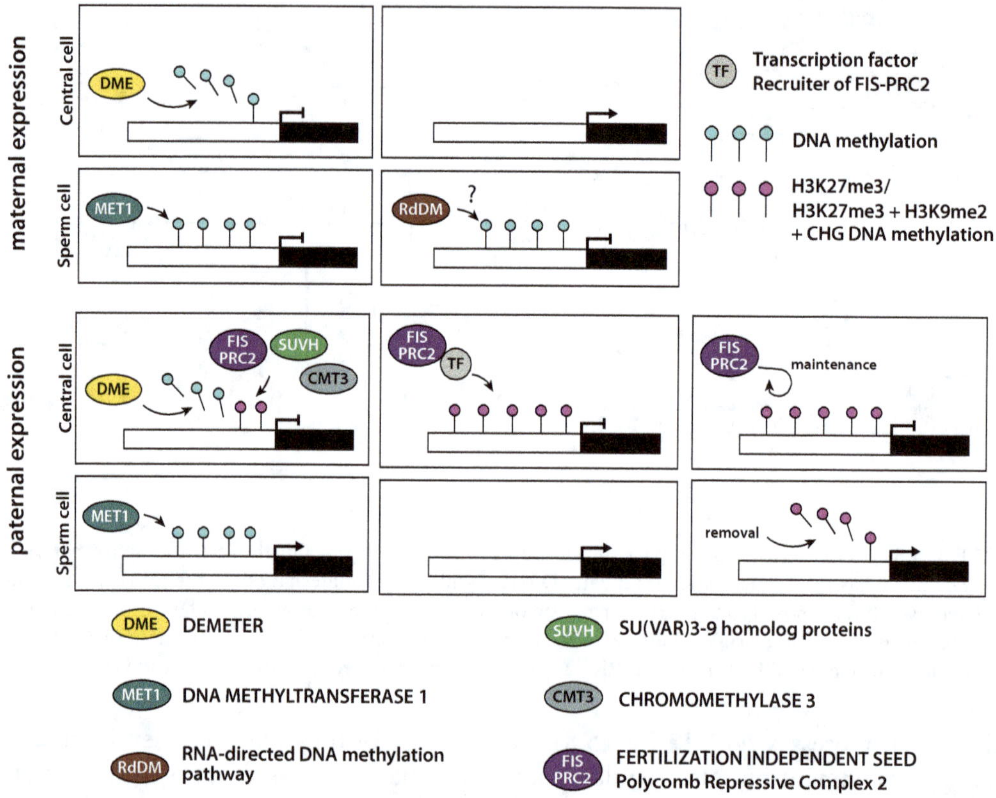

Fig. 5.9 Different models for the regulation of genomic imprinting in the endosperm. The events proposed to set gametic imprints in central cell and sperm are shown. For maternally expressed genes, the establishment of differential DNA methylation imprints, through demethylation or specific *de novo* methylation, is important. For paternally expressed genes, three scenarios involving PRC2 are shown. See text for details. (From Batista and Köhler 2020)

neither involve DNA nor histone methylation. This is particularly true for imprinted genes in the embryo, where none of 12 tested genes appears to be regulated by DNA methylation and only three were affected by mutants affecting PRC2 (Raissig et al. 2013).

There are also several cases of imprinted genes expressed in maize or *A. thaliana* endosperm carrying a DMR that, however, is not inherited from the parents but only established after fertilization. Differential DNA methylation does, therefore, not act as a gametic imprint in these cases.

That imprinting can be very complex was also shown through more detailed studies of the *MEA* locus. The expression of the maternal *MEA* allele depends on demethylation by DME and, after fertilization, the paternal allele is repressed by the maternally produced FIS-PRC2. The latter was concluded from several studies in which a paternal reporter gene driven by the *MEA* promoter became active in crosses to *mea* mothers. However, the analysis of the endogenous gene subsequently showed that the derepression of the paternal allele was minimal and *MEA* still showed a strongly biased allelic expression. Furthermore, imprinting could be conferred to a reporter gene using only 200 bp of the *MEA* promoter, which is completely free of any DNA methylation in the gametes. This was the first report of a DNA methylation-independent ICE, which were subsequently also found in mammals. As neither PRC2 nor DNA methylation is required for imprinted *MEA* expression, the primary imprints remain elusive (Wöhrmann et al. 2012). It was proposed that, similar to *PHE1*, chromatin loops may form depending on the DNA methylation status of flanking regions, thereby preventing access of imprinting factors that would establish the yet unknown gametic imprint.

Finally, the identification of *PHE1* as a target of FIS-PRC2 revealed the existence of an extensive regulatory cascade controlling seed development that is based on genomic imprinting. The maternally expressed imprinted genes *MEA* and *FIS2* control the paternal expression of *PHE1*, a transcription factor that, in turn, regulates a large number of paternally expressed imprinted genes in the endosperm. Thus, epigenetic gene regulation plays a major role in seed development.

5.5 Evolution of Genomic Imprinting

Several hypotheses have been proposed to explain the evolution of genomic imprinting. Among these, David Haig's parental conflict theory by far attracted the most attention from biologists working on imprinting. Haig tried to find an explanation why imprinting evolved in such different organisms as mammals and seed plants (Haig and Westoby 1989). He realized that both share a "placental habit", i.e., the development of seeds and fetuses depends solely on resources provided by the mother. Indeed, both the mammalian placenta and the endosperm of seed plants are major tissues with imprinted gene expression and play a central role in providing nutrients to the next generation. In contrast, the father contributes little to support the developing offspring. In polygamous species, this leads to a parental conflict between the paternal and maternal genomes in the progeny because of different kinship relationships. The mother is equally related to all her offspring and (epi)genotypes are favored that lead to a uniform distribution of resources to all her offspring over her lifetime. In contrast, the paternal genome is not the same in all the progeny, such that (epi)genotypes are favored that lead to an increased acquisition of nutrients at the expense of non-related siblings. This parental conflict will lead to the biased expression of genes involved in the acquisition of nutrients from the mother, and thus the growth of the seed or fetus. This theory makes clear evolutionary predictions: First, paternally expressed genes should promote growth of the offspring and second, maternally expressed genes should reduce it.

There is substantial evidence in support of the parental conflict theory (Pires and Grossniklaus 2014). For instance, the endosperm overproliferates in crosses of parents with different ploidy that increase the number of paternal genomes, while it proliferates less and differentiates earlier in seeds with an excess of maternal genomes (Haig and Westoby 1989). Similarly, androgenetic mouse embryos form a large trophoblast, from which the embryonic part of the placenta forms, while gynogenetic embryos produce an underdeveloped trophoblast. The parental conflict theory is also supported by its phylogenetic distribution in vertebrates. Genes that are imprinted in mammals are biallelically expressed in fish and birds, in which the amount of resources deposited in the egg is determined by the mother prior to fertilization, such that the paternal genome in zygotic tissues cannot influence nutrient acquisition from the mother. This is also true for egg-laying mammals like the platypus and echidna but not for marsupials. In the latter, much of the offspring's development occurs in the pouch, which from the point-of-view of the parental conflict theory is equivalent to the uterus in eutherian mammals or the seed in plants.

Intriguingly, the predictions of the parental conflict theory also play out at the level of individual genes. For instance, *MEA* is a maternally expressed gene and is predicted to restrict growth. Indeed, seeds developing from *mea* mutant gametes show overproliferation in both the embryo and endosperm before they abort (Grossniklaus et al. 1998). Similarly, the paternally expressed *Igf2* gene is expected to promote fetal growth in the mouse and, consistent with the theory, mutant pubs have a 40% reduction in birth weight compared to their wild-type siblings. In contrast, embryos lacking the maternally expressed *Igf2r* gene, encoding the receptor for Igf2, show overgrowth before they die. Most interestingly, double mutants are normal in size and are viable. In fact, Igf2r is a receptor that binds the growth factor Igf2 but targets it to the lysosome, preventing its growth-promoting action. Thus, Igf2r is a kind of decoy receptor expressed from the maternal genome to remove the paternally produced Igf2 growth factor: a molecular reflection of the tug-of-war between maternal and paternal genomes.

The parental conflict theory is very attractive as it explains the evolution of imprinting in both mammals and plants. To my knowledge, the phenotypes of all imprinted genes that have a function related to growth, either during embryogenesis or, in mice, also postnatally, e.g., by controlling feeding behavior, are consistent with the parental conflict theory. This includes some of the clinical features of patients with syndromes associated with imprinted gene clusters, which often affect birth weight or postnatal feeding behavior. However, not all imprinted genes regulate growth and for these, the parent conflict theory does not apply in an obvious manner. More than a dozen other theories for the evolution of imprinting have been proposed (Spencer and Clark 2014). Depending on the organism and the function of the imprinted gene, distinct selective forces may have driven the evolution of genomic imprinting. However, most alternative theories do not explain the peculiar phylogenetic distribution of genomic imprinting, i.e., its evolution in mammals and seed plants.

Take-Home Messages

- Genomic imprinting is a paradigm of epigenetic gene regulation that evolved independently in seed plants and mammals.
- Genomic imprinting leads to parent-of-origin dependent gene expression, rendering parental genomes non-equivalent, and is not related to the sex of the individual.
- Gametic or primary imprints are differential epigenetic marks that are acquired during gametogenesis and lead to allelically biased gene expression after fertilization.
- Gametic imprints have to be reset every generation in the germline according to the sex of the individual. In mammals, they withstand genome-wide reprogramming in the zygote, associated with a wave of DNA demethylation and remethylation.

- In mammals, imprinted genes typically occur in clusters that contain at least one maternally and one paternally expressed gene. Most clusters also contain a long non-coding RNA that is involved in imprinting regulation.
- Parental-specific expression in a cluster is regulated by imprinting control elements (ICEs) that are differentially marked, most often by DNA methylation imprints or, more rarely, by histone modifications that serve as gametic imprints.
- ICEs are long-range regulatory elements that control the expression of multiple genes in the cluster. Deletion of ICEs affects expression on the chromosome carrying the imprint.
- In both mammals and seed plants, genomic imprinting is regulated by DNA methylation and PRC2, although the former has a more predominant role in mammals. Differential DNA methylation can lead to topological changes, such as the formation of chromatin loops, that control imprinted expression.
- Although non-coding RNAs seem to play a role in the regulation of many imprinted gene clusters in mammals, the clusters are diverse and regulated in distinct ways, e.g., by modifying insulator elements.
- In plants, differential DNA methylation is associated with about 30% of the maternally expressed imprinted genes, the mechanisms regulating the rest are unknown. The majority of paternally expressed genes are controlled by PRC2, DNA methylation, or a combination thereof.
- Imprinted genes play an important role in development and behavior and, in agreement with the parental conflict theory for the evolution of imprinting, many of them regulate growth both in mammals and seed plants.

References

Barlow DP, Bartolomei MS (2014) Genomic imprinting in mammals. Cold Spring Harb Perspect Biol 6:a018382. https://doi.org/10.1101/cshperspect.a018382

Barlow DP, Stöger R, Herrmann BG, Saito K, Schweifer N (1991) The mouse insulin-like growth factor type-2 receptor is imprinted and closely linked to the *Tme* locus. Nature 349:84–87

Baroux C, Spillane C, Grossniklaus U (2002) Genomic imprinting during seed development. Adv Genet 46:165–214. https://doi.org/10.1016/s0065-2660(02)46007-5

Bartolomei MS, Zemel S, Tilghman SM (1991) Parental imprinting of the mouse *H19* gene. Nature 351:153–155

Batista RA, Köhler C (2020) Genomic imprinting in plants – revisiting existing models. Genes Dev 34:24–36. https://doi.org/10.1101/gad.332924.119

Choi Y, Gehring M, Johnson L, Hannon M, Harada JJ, Goldberg RB, Jacobsen SE, Fischer RL (2002) DEMETER, a DNA glycosylase domain protein, is required for endosperm gene imprinting and seed viability in *Arabidopsis*. Cell 110:33–42. https://doi.org/10.1016/S0092-8674(02)00807-3

Cooper DW, VandeBerg JL, Sharman GB, Poole WE (1971) Phosphoglycerate kinase polymorphism in kangaroos provides further evidence for paternal X inactivation. Nat New Biol 230:155–157. https://doi.org/10.1038/newbio230155a0

DeChiara TM, Robertson EJ, Efstratiadis A (1991) Parental imprinting of the mouse insulin-like growth factor II gene. Cell 64:849–859

Evans MMS, Grossniklaus U (2008) The maize megagametophyte. In: Bennetzen J, Hake S (eds) Handbook of maize: its biology. Springer, New York, pp 79–104. https://www.springer.com/gp/book/9780387794174

Grossniklaus U, Vielle-Calzada J-P, Hoeppner MA, Gagliano WB (1998) Maternal control of embryogenesis by *MEDEA*, a *Polycomb*-group gene in *Arabidopsis*. Science 280:446–450

Haig D, Westoby M (1989) Parent-specific gene expression and the triploid endosperm. Am Nat 134:147–155. https://doi.org/10.1086/284971

Hsieh TF, Ibarra CA, Silva P, Zemach A, Eshed-Williams L, Fischer RL, Zilberman D (2009) Genome-wide demethylation of *Arabidopsis* endosperm. Science 324:1451–1454. https://doi.org/10.1126/science.1172417

Inoue A, Jiang L, Lu F, Suzuki T, Zhang Y (2017) Maternal H3K27me3 controls DNA methylation-independent imprinting. Nature 547:419–424. https://doi.org/10.1038/nature23262

Iqbal K, Jin SG, Pfeifer GP, Szabó PE (2011) Reprogramming of the paternal genome upon fertilization involves genome-wide oxidation of 5-methylcytosine. Proc Natl Acad Sci U S A 108:3642–3647. https://doi.org/10.1073/pnas.1014033108

Kermicle JL (1970) Dependance of the *R*-mottled aleurone phenotype in maize on mode of sexual transmission. Genetics 66:69–85. PMID: 17248508

Köhler C, Page DR, Gagliardini V, Grossniklaus U (2005) The *Arabidopsis thaliana* MEDEA *Polycomb* group protein controls expression of *PHERES1* by parental imprinting. Nat Genet 37:28–30. https://doi.org/10.1038/ng1495

Loppin B, Oakey RJ (2009) Genomic imprinting in Singapore. EMBO Rep 10:222–227. https://doi.org/10.1038/embor.2009.20

McGrath J, Solter D (1984) Completion of mouse embryogenesis requires both the maternal and paternal genomes. Cell 37:179–183. https://doi.org/10.1016/0092-8674(84)90313-1

Messing J, Grossniklaus U (1999) Genomic imprinting in plants. Results Probl Cell Differ 25:23–40. https://doi.org/10.1007/978-3-540-69111-2_2

Metz CW (1938) Chromosome behavior, inheritance and sex determination in *Sciara*. Am Nat 72:485–520. https://www.jstor.org/stable/2457532

Monk D, Mackay DJG, Eggermann T, Maher ER, Riccio A (2019) Genomic imprinting disorders: lessons on how genome, epigenome and environment interact. Nat Rev Genet 20:235–248. https://doi.org/10.1038/s41576-018-0092-0

Oakey RJ, Beechy CV (2002) Imprinted gene: identification by chromosome rearrangements and post-genomic strategies. Trends Genet 7:359–366. https://doi.org/10.1016/S0168-9525(02)02708-7

Pires ND, Grossniklaus U (2014) Different yet similar: evolution of imprinting in flowering plants and mammals. F1000Prime Rep 6:63. https://doi.org/10.12703/P6-63

Raissig MT, Bemer M, Baroux C, Grossniklaus U (2013) Genomic imprinting in the *Arabidopsis* embryo is partly regulated by PRC2. PLoS Genet 9:e1003862. https://doi.org/10.1371/journal.pgen.1003862

Reik W, Walter J (2001) Genomic imprinting: parental influence on the genome. Nat Rev Genet 2:21–32. https://doi.org/10.1038/35047554

Sánchez L (2014) Sex-determining mechanisms in insects based on imprinting and elimination of chromosomes. Sex Dev 8:83–103. https://doi.org/10.1159/000356709

SanMiguel JM, Bartolomei MS (2018) DNA methylation dynamics of genomic imprinting in mouse development. Biol Reprod 99:252–262. https://doi.org/10.1093/biolre/ioy036

Spencer HG, Clark AG (2014) Non-conflict theories for the evolution of genomic imprinting. Heredity 113:112–118. https://doi.org/10.1038/hdy.2013.129

Stewart KR, Veselovska L, Kelsey G (2016) Establishment and functions of DNA methylation in the germline. Epigenomics 8:1399–1413. https://doi.org/10.2217/epi-2016-0056

Surani MA, Barton SC, Norris ML (1984) Development of reconstituted mouse eggs suggests imprinting of the genome during gametogenesis. Nature 308:548–550. https://doi.org/10.1038/308548a0

Vielle-Calzada J-P, Thomas J, Spillane C, Coluccio A, Hoeppner MA, Grossniklaus U (1999) Maintenance of genomic imprinting at the *Arabidopsis medea* locus requires zygotic *DDM1* activity. Genes Dev 13:2971–2982

Walbot V, Evans MMS (2003) Unique features of the plant life cycle and their consequences. Nat Rev Genet 4:369–379. https://doi.org/10.1038/nrg1064

Wöhrmann HJ, Gagliardini V, Raissig MT, Wehrle W, Arand J, Schmidt A, Tierling S, Page DR, Schöb H, Walter J, Grossniklaus U (2012) Identification of a DNA methylation-independent imprinting control region at the *Arabidopsis MEDEA* locus. Genes Dev 26:1837–1850. https://doi.org/10.1101/gad.195123.112

Wyder S, Raissig MT, Grossniklaus U (2019) Consistent reanalysis of genome-wide imprinting studies in plants using generalized linear models increases concordance across datasets. Sci Rep 9:1320. https://doi.org/10.1038/s41598-018-36768-4

Zoghbi HY, Beaudet AL (2016) Epigenetics and human disease. Cold Spring Harb Perspect Biol 8:a019497. https://doi.org/10.1101/cshperspect.a019497

6

Mechanisms of the RNA-Based Gene Silencing

Contents

What You Will Learn in This Chapter

Although epigenetic states are typically associated with DNA-methylation and posttranslational histone modifications, RNAs often play an important role in their regulation. Specific examples have already been discussed in the context of dosage compensation (see book ► Chap. 4 of Wutz) and genomic imprinting (see book ► Chap. 5 of Grossniklaus). In this Chapter, we will take a closer look at a particular class of RNAs implicated in gene silencing. Although the focus will lie on RNA-based silencing mechanisms in plants, many of its components, such as RNase III-related DICERLIKE endonucleases or small RNA-binding ARGONAUTE proteins, are conserved in animals, plants, and fungi. On the one hand, small RNAs are involved in post-transcriptional silencing by targeting mRNAs for degradation or inhibiting their translation, a feature that has been exploited for large-scale genetic screens. On the other hand, they also play a central role in transcriptional gene silencing, for instance in the repression of transposable elements across a wide variety of organisms. In plants, this involves a complex system whereby small RNAs derived from transposons and repeats direct DNA-methylation and repressive histone modifications in a sequence-specific manner. Recent results link this so-called RNA-dependent DNA-methylation to paramutation, a classical epigenetic phenomenon where one allele directs a heritable epigenetic change in another.

6.1 The Unusual Behavior of Transgenes Led to the Discovery of Novel RNA-Based Silencing Mechanisms

With the establishment of transformation technologies and their increasing use in plants in the late 1980s, it became apparent that the expression behavior of transgenes was hard to predict. While transgenics that did not show the expected expression of the transgene were initially discarded, they soon became an object of interest. One of the most striking examples was the characterization of transgenic petunia (*Petunia hybrida*) overexpressing the chalcone synthase (*CHS*) gene by Carolyn Napoli and Richard Jorgensen (Napoli et al. 1990). *CHS* encodes the rate-limiting enzyme of anthocyanin biosynthesis, which leads to the purple pigmentation of the flowers (◘ Fig. 6.1). Unexpectedly, instead of producing deep purple flowers due to *CHS* overexpression, many transgenics produced white flowers or flowers with a combination of purple and white tissues. Molecular analyses showed that the transgene had an effect on the endogenous *CHS* gene in *trans*, and that the transgene and the endogenous *CHS* gene were coordinately repressed (◘ Fig. 6.1). This phenomenon was termed "co-suppression". Subsequently, a similar observation was reported in the fungus *Neurospora crassa*, where the introduction of two transgenes of the carotenoid biosynthesis pathway led to a reduction in the expression of the endogenous gene, referred to as "quelling" (Romano and Macino 1992). RNA-based silencing was also discovered in animals, where the injection of RNA being either sense or antisense to the *par-1* mRNA in the nematode *Caenorhabditis elegans* led to degradation of the *par-1* transcripts (Guo and Kemphues 1995). Unlike the chromatin-based mechanisms described in earlier chapters, in these cases, gene silencing occurred at the post-transcriptional level (PTGS). Such cytoplasmic mechanisms can also contribute to the establishment of heritable epigenetic states as the RNAs involved are sometimes amplified in a self-sustaining fashion. Other RNA-based gene silencing

Fig. 6.1 Discovery of RNA-based gene silencing in petunia. **a** The wild type produces purple flowers due to anthocyanin production requiring the *CHS* biosynthetic gene (left panel). Introduction of a *CHS* transgene leads to a variety of phenotypes, including completely white flowers (three panels on the right). (From Stam et al. 1997). **b** Some of these phenotypes are unstable, producing revertant sectors (left panel). RNAse protection assays showed that both transgene (TG) and endogenous *CHS* gene (EG) are expressed in the purple sectors but RNA levels of both TG and EG are strongly reduced in the white sectors (right panels). Three flowers per sector were analyzed; M, marker lane. (From Napoli et al. 1990)

phenomena in plants (Matzke et al. 1989) and animals (Bestor et al. 1994) occur in the nucleus and involve transcriptional gene silencing (TGS) by DNA methylation and chromatin modifications. It was soon clear that the various gene silencing phenomena were based on homology, i.e., showed sequence-specificity, and models involving DNA-DNA, DNA-RNA, and RNA-RNA pairing were proposed.

The seminal experiments of Fire and Mello (Fire et al. 1998) in *C. elegans* showed that not single-stranded but rather double-stranded RNA (dsRNA) was responsible for silencing, providing a unifying explanation for co-suppression, quelling, and other PTGS phenomena, now referred to as RNA interference (RNAi). In both plants and nematodes, gene silencing was heritable, suggesting a stable silencing intermediate. Hamilton and Baulcombe (1999) showed that this was not the full-length dsRNA but rather short fragments of roughly 25 nt in length, so-called short interfering RNAs (siRNAs). Similar 21–23 nt long siRNAs were then also found to be involved in RNAi in *Drosophila melanogaster* and *C. elegans.* The demonstration that chemically synthesized 21–22 nt dsRNAs can downregulate target mRNAs also in mammalian cells, opened completely new avenues for functional studies in cell culture. These discoveries did not only stimulate intensive research into the underlying mechanisms but also led to a wide range of applications (Sen and Blau 2006).

6.1.1 Conserved Components of RNA-Based Silencing Mechanisms

Given that siRNAs are involved in diverse silencing mechanisms in animals and plants, it is not surprising that they also use similar proteins that mediate the function of siRNAs. As the mechanisms are diverse, only the two key components are introduced here: the enzymes responsible for dicing the dsRNA into the siRNAs [DICERLIKE (DCL) proteins] and the ones involved in slicing the target mRNAs [ARGONAUTE (AGO) proteins]. DCL and AGO proteins are central to essentially all RNA-based silencing mechanisms.

DCL proteins, named after the founding member Dicer, were identified in *D. melanogaster* through a candidate gene approach testing different RNase III family members for the ability to cleave dsRNA into siRNAs (Bernstein et al. 2001). DCL proteins have a conserved structure, with two RNase III domains that cleave the dsRNA, a PAZ domain binding the 3'end of the dsRNA, and a helicase domain that unwinds the dsRNA duplexes (◘ Fig. 6.2). In some species, e.g., the model plant *Arabidopsis thaliana*, the family diversified and different members have specialized functions.

Another family of proteins involved in various silencing pathways are the ARGONAUTE (AGO) proteins, which are the catalytic subunit of an RNA-induced silencing complex (RISC) that slices the target mRNAs. Like DCL proteins, AGOs have a PAZ domain that binds the 3'end of dsRNAs, in this case the siRNAs produced by DCL proteins, which can base-pair with complementary target mRNAs. Through their PIWI domain, AGOs can slice the target mRNA complementary to the siRNA (◘ Fig. 6.3). However, cleaving the mRNA is not the only outcome of base-pairing with the siRNA; depending on the silencing pathway and type of RISC, translation of the mRNA can be blocked instead. AGO proteins can also be incorporated into another complex, called RNA-induced transcriptional silencing (RITS) complex, which is targeted to homologous DNA regions. The RITS complex then recruits enzymes that mediate DNA methylation and/or histone modifications which are associated with TGS.

6.2 Post-Transcriptional Gene Silencing (PTGS)

There is a wide variety of small RNAs that are involved in targeting mRNAs for degradation or inhibiting translation at the post-transcriptional level (PTGS). Others are involved in the sequence-specific silencing of genes at the transcriptional level (TGS) by establishing repressive chromatin (Pikaard and Mittelsten Scheid 2014). These small RNAs can be derived from transgenes or viruses (siRNAs) but many of them are endogenously produced during the life cycle of the plant, including microRNAs (miRNAs), hairpin siRNAs (hp-siRNA), trans-acting siRNAs (tasiRNAs),

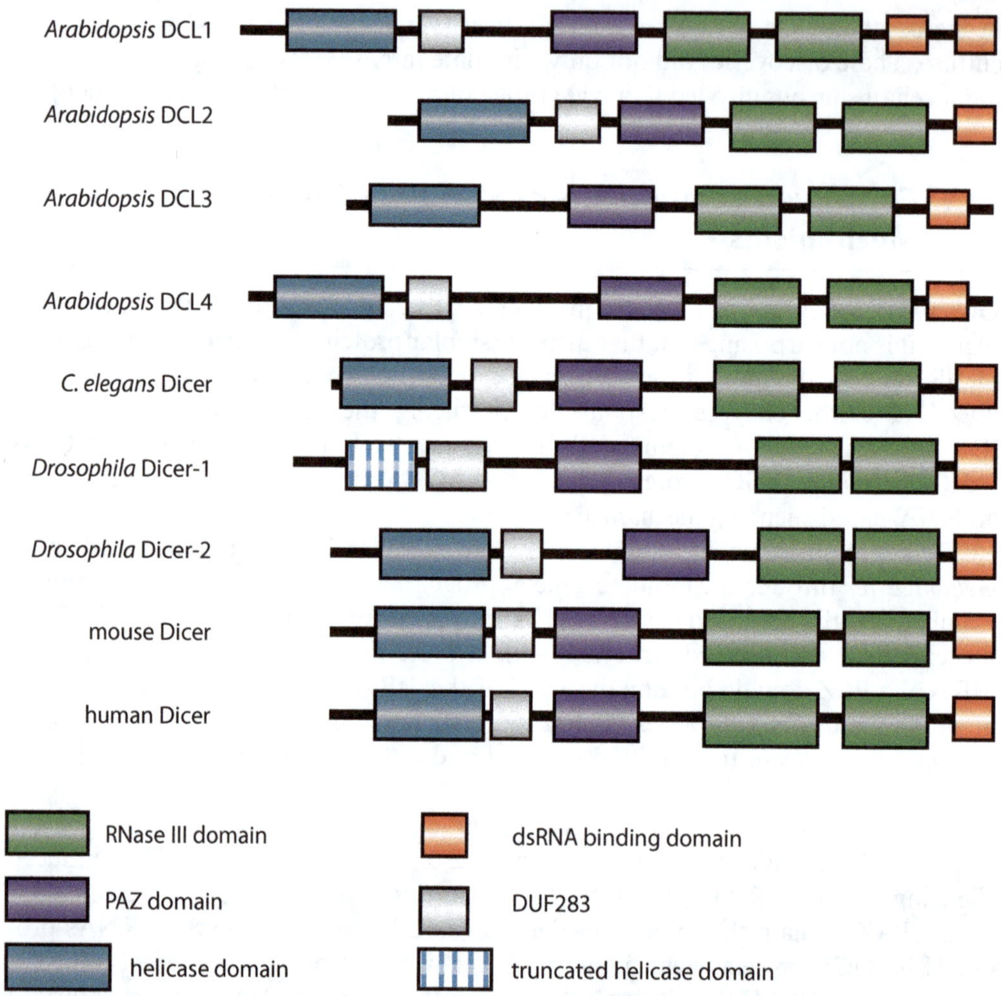

Fig. 6.2 Domain organization of DICERLIKE (DCL) proteins from various eukaryotes. DCL proteins are responsible for dicing dsRNAs of diverse origin, e.g., derived from transgenes, endogenous genes, or viruses, into siRNAs. They are also crucial for the biosynthesis of microRNAs (miRNAs), regulatory small RNAs produced from stem-loop structures formed by transcripts of endogenous loci. The RNase III domains are responsible for dicing, while other domains are involved in binding the target dsRNA. (From Johanson et al. 2013)

phased siRNAs (phasi RNAs) produced during anther development, and epigenetically activated siRNA (easiRNA) that accumulate in the sperm cells (Fig. 6.4). Here, we will look at the biogenesis and function of only a few selected types of small RNAs (Borges and Martienssen 2015).

6.2.1 The Biogenesis and Function of microRNAs

In both animals and plants, miRNAs play important roles in development by limiting the function of their target mRNAs to specific cells or tissues through their degradation or translational repression. Thus, they regulate gene expression at the post-transcriptional level and have an effect as long as they are present in the same

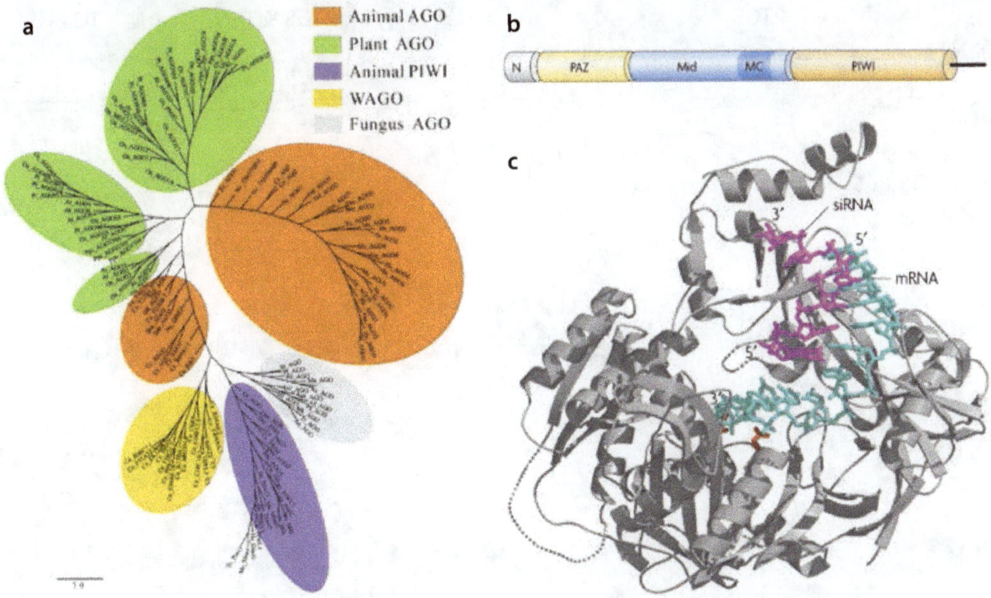

Fig. 6.3 Phylogenetic tree and domain structure of ARGONAUTE (AGO) proteins. **a** Unrooted phylogenetic tree of 135 AGO proteins. The AGO proteins fall into three major clades: AGO proteins from plants, animals, and fungi that bind miRNAs or siRNAs, PIWI proteins which bind to Piwi-interacting RNAs (piRNAs) expressed in the germline of animals only, and the worm-specific WAGO clade. In some species, this family is highly diversified, e.g., in *A. thaliana* with 10 and *C. elegans* with 25 members, respectively. (From Wei et al. 2013). **b** Domain structure of AGO proteins with a PAZ domain that binds siRNAs, the PIWI domain that cleaves the target mRNA, and the cap-binding-like domain MC that lies within the Mid-domain connecting the PAZ and PIWI domains. (From Hutvagner and Simard 2008). **c** Crystal structure of an AGO protein, showing the 3'end of the siRNA in purple and part of the mRNA in turquoise; active residues of the PIWI domain are shown in red. (From Song et al. 2004)

cell as the target mRNA. Because they do not result in a heritable change of gene expression, they are often not considered epigenetic regulators. An epigenetic change in gene expression persists even if the initial trigger, be it developmental or environmental, is gone (see also book ► Chap. 3 of Paro). Nevertheless, because their biogenesis and function are similar to epigenetically acting, short RNAs whose level is sustained, e.g. by RNA-dependent RNA polymerases (RDRs) that generate dsRNAs in a feed-forward loop, they are also discussed here.

miRNAs are derived from RNA polymerase II (Pol II) transcribed genes that do not encode proteins, but they can also originate from sequences in long non-coding RNAs or introns of protein-coding genes. These precursor RNAs have stretches of self-complementarity and can form imperfect RNA hairpins or stem-loop structures, which are then processed into the miRNAs. In mammals, these pri-miRNAs are cleaved by the RNase III-type endonuclease Drosha on one side of the loop in the nucleus. The resulting partially processed pre-miRNA is then transported to the cytoplasm where Dicer cuts the stem 22 nt away from the Drosha cleavage site to generate an RNA duplex consisting of the miRNA and its complementary strand. In plants, both cleavages are made by DCL1 in the nucleus and result in 21 nt long miRNA duplexes (◘ Fig. 6.4). Before these are exported to the cytoplasm, they are processed by HUA ENHANCER1 (HEN1), which methylates the 3′-terminal ribose of each RNA strand to protect the miRNA from degradation. The miRNA is then

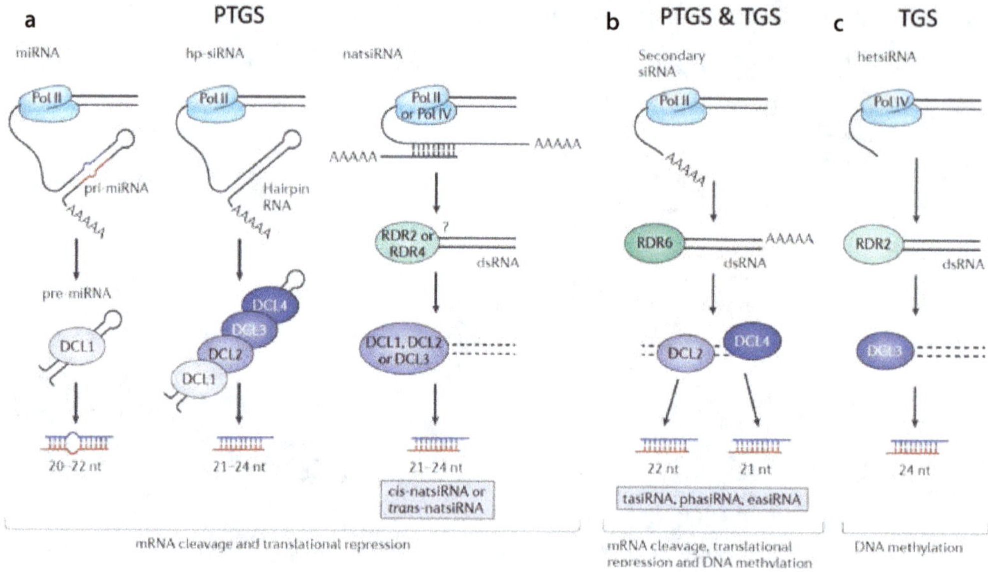

Fig. 6.4 Different small RNA biogenesis pathways in plants. **a** Pathways involved in PTGS: transcripts from miRNA genes form hairpin-like structures (pri-miRNA) that are diced by DLC1 into shorter stem-loop structures (pre-miRNA) and then into mature miRNA duplexes. Longer hairpin structures can originate from inverted repeats and are processed by all DCL proteins into hp-siRNAs. They may represent an early stage in the evolution of miRNAs. natsiRNAs are derived from dsRNA produced by overlapping transcription (*cis*-natsiRNAs) or from transcripts with regions of high complementary originating from distant loci (*trans*-natsiRNAs). RNA-dependent RNA polymerases (RDRs) are likely involved in extending the dsRNA, which is diced by various DCLs. **b** Secondary siRNAs act on transcripts distinct from those that generated them and are involved in both PTGS and TGS. The transcripts come from non-coding loci, protein-coding genes, or transposons and are converted into dsRNA by RDR6 and processed by DCL2 and DCL4 into various classes of siRNAs. The tasiRNAs, for instance, are derived from non-coding transcripts of specific *TAS* genes (*TAS1* to *TAS4* in *A. thaliana*), which are targeted by specific miRNAs that mediate their cleavage (e.g., miR173-AGO1 for *TAS1*/*TAS2*; mir390-AGO7 for *TAS3*). The 3′ cleavage product is converted into dsRNA by RDR6, and then sequentially diced by DCL4 into 21 nt long tasiRNAs that are phased relative to the miRNA cleavage site. Thus, tasiRNAs have defined sequences and function like miRNAs to slice or inhibit their target mRNAs. **c** The heterochromatic siRNAs (hetsiRNAs) originate from transposons and repeats, often located in pericentromeric regions. They depend on transcription by RNA polymerase IV (Pol IV), dsRNA synthesis by RDR2, and dicing into 24 nt long siRNAs by DCL3. hetsiRNAs mediate RNA-dependent DNA methylation (RdDM). (From Borges and Martienssen 2015)

bound by *A. thaliana* AGO1 (sometimes AGO7/AGO10) to form a RISC complex that slices the target mRNA or blocks its translation. In plants, the miRNA typically shows near-perfect complementarity to the target mRNA, leading to slicing in the majority of cases. In contrast, animal miRNAs rarely have perfect complementarity to the mRNA and more often cause translational inhibition. Also in plants, some miRNA block translation in combination with slicing of the transcript. Many plant miRNAs target transcription factors and are often conserved during evolution, indicating important regulatory roles during development. Although the coding sequences for these transcription factors have diverged considerably, the target sites on the mRNAs matching the miRNAs are nearly invariant. Other miRNAs are evolutionary young and often specific to a cell lineage. In *A. thaliana*, these lineage-specific miRNAs are sometimes processed by DCL4, while DCL1 is required for the vast majority of miRNAs.

6.2.2 Genome Defense by siRNA-Mediated Silencing

An important class of small RNAs are the siRNAs that induce PTGS to defend the genome against viruses and microbes (Pikaard and Mittelsten Scheid 2014). Most plant viruses have a ssRNA genome whose replication depends on an endogenous or virus-encoded RDR to produce dsRNA intermediates, which are substrates for DCL proteins. In addition, viruses with dsRNA genomes are direct substrates for dicing as are dsRNA intermediates formed during the life cycle of ssDNA viruses. The silencing of transgenes is likely observed due to the activity of similar genome defense pathways. Transgenes often integrate as multiple copies that can have opposite orientations. Transcription through inverted transgenes can lead to hairpin-like RNA structures that are substrates for DCL proteins. Transgenes may also integrate next to strong promoters pointing in the opposite direction, leading to transcription of both strands and, thus, dsRNA formation. Finally, other aberrant types of transgene-derived mRNAs can serve as a substrate for an RDR protein, generating dsRNA. PTGS of viruses and transgenes typically involves 21–22 nt long siRNAs generated primarily by DCL2 and DCL4 (◘ Fig. 6.4). The 21 nt long siRNAs guide the cleavage of the mRNA by AGO1, which generates a 3′ fragment of the mRNA that lacks a 7-methylguanosine cap at its 5’end. mRNAs without a cap are eliminated rapidly by exonuclease-mediated degradation. The association of 22 nt siRNAs with AGO1 couples target mRNA cleavage to the recruitment of RDR6, such that the 3′ cleavage fragment is transcribed into dsRNA, which is then diced again by DCL4 to make 21 nt long secondary siRNAs. DCL3 can also dice dsRNA derived from viruses or transgenes, generating 24 nt siRNAs (◘ Fig. 6.4). However, these lead to TGS through DNA and chromatin modifications at loci with complementarity to the siRNAs. Taken together, these diverse aspects induced by siRNAs mount a very powerful response to invading nucleic acids. Not surprisingly, viruses developed a variety of countermeasure and many encode suppressors of gene silencing in their genomes (Jiang et al. 2012).

The discovery of RNAi and its underlying mechanisms, many of which are conserved across organisms, have opened unprecedented opportunities for applications. For instance, the discovery that synthetic dsDNA can be used in tissue culture cells to induce RNAi allowed genome-wide screens to address gene function in cultured cells. This has been instrumental not only for basic research but especially for drug discovery in the pharmaceutical industry (Hokaiwado et al. 2008). In plants, miRNAs have been used to address the function of genes or gene families by introducing artificial miRNAs that target transcripts derived from one or multiple loci. Also, siRNAs derived from transgenic hairpin constructs that target a specific gene of interest have been extensively used to study gene function. Moreover, viral genomes have been engineered to encode plant genes of interest. A viral infection thereby leads to the production of siRNAs against the endogenous plant gene, allowing functional studies. This so-called virus-induced gene silencing (VIGS) is particularly useful to study the function of genes in non-model organisms, where no efficient transgenesis methods are available but infections using such engineered viruses are possible.

Similar approaches have led to biotechnological applications. One of the most successful advancements was the development of virus-resistant plants by exploiting

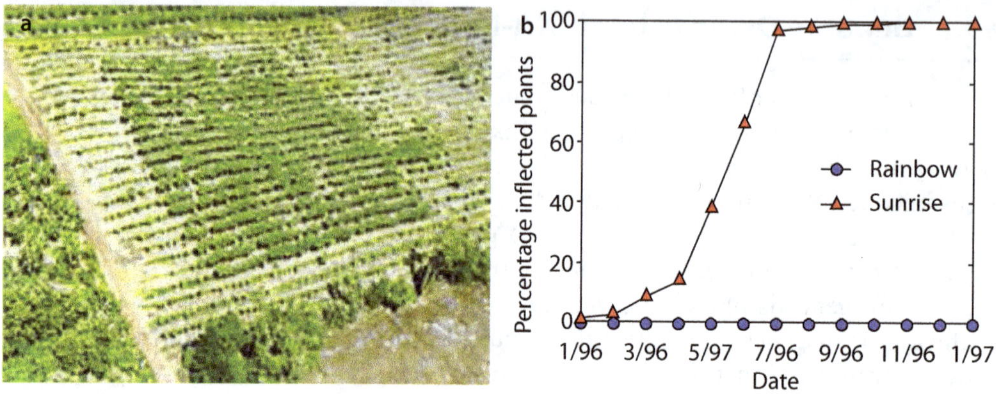

Fig. 6.5 Virus-resistant, transgenic papaya based on RNAi directed against the coat protein of papaya ringspot virus. **a** Field trial showing a block of resistant papaya plants of the transgenic Rainbow cultivar surrounded by susceptible non-transgenic plants of the Sunrise cultivar. **b** All transgenic Rainbow plants were resistant to the virus while all non-transgenic Sunrise plants became infected within a few months of the field trial. (From Gonsalves et al. 2004)

the siRNA-based defense mechanism of the plant and targeting it specifically to the virus. For instance, the Hawaiian production of papaya was severely threatened by papaya ringspot virus, which devastated the plantations in the 1990s. The introduction of a genetically modified virus-resistant papaya variety in 1998 saved the papaya growers and is the basis of current papaya production in the USA (Fig. 6.5). A similar approach has since been used to protect commercially grown summer squash (Gottula and Fuchs 2009).

6.3 Transcriptional Gene Silencing (TGS)

Transgene silencing does not only involve siRNAs at the post-transcriptional level but also comprises DNA methylation. The two phenomena were linked when it was discovered that homology-dependent DNA methylation also involved siRNAs (Mette et al. 2000) and some mutants of *A. thaliana* affected both PTGS and TGS. Furthermore, the silencing of transposable elements and repeats is crucial for the maintenance of genome integrity. A failure to keep transposons in check results in hybrid dysgenesis in *D. melanogaster*, where massive transposition in the germline leads to sterility, a high mutation rate, and chromosome breakage. In metazoans, transposons are silenced through the action of Piwi-interacting RNAs (piRNAs), which are generated from long Pol II transcripts that are derived from piRNA clusters, loci composed of transposons of diverse families or fragments thereof. Thus, piRNAs are complementary to many dispersed transposons. Their production is independent of DCL proteins but piRNAs bind to AGO-like proteins of the PIWI clade (Fig. 6.3), which are essential to their biogenesis. Primary piRNAs can be amplified through a so-called ping-pong cycle and eventually lead to silencing of the transposons by recruiting the machinery to establish H3K9me3 and, depending on the organism, *de novo* DNA methyltransferases (Ernst et al. 2017).

The silencing of endogenous repeats, such as those present at centromeres, was also found to rely on RNA-based mechanisms, for instance in the formation of centromeric heterochromatin in *A. thaliana* or the yeast *Schizosaccharomyces pombe* (Martienssen and Moazed 2015). Here, however, we focus on RNA-dependent DNA-methylation (RdDM), an RNA-based mechanism primarily targeting endogenous transposons and repeats, but also transgenes, in plants (Matzke and Mosher 2014).

RdDM was initially discovered in plants that were infected with a viroid (Wassenegger et al. 1994). It was later shown to also occur at transcriptionally silenced transgenes and rely on siRNAs and proteins of the RNAi machinery. RdDM is a very complex mechanism, involving many factors that were identified in genetic screens for mutants that relieve transgene silencing. In addition, the biochemical isolation of proteins interacting with known factors by co-immunoprecipitation contributed to complete the picture. Taken together, these studies revealed that RdDM relies on two specialized RNA polymerase complexes that are derived from Pol II, called Pol IV and Pol V. Conceptually, the RdDM mechanism can be separated into three major phases (◘ Fig. 6.6) (Matzke and Mosher 2014).

First, Pol IV is recruited to target loci by SAWADEE HOMEODOMAIN HOMOLOGUE1 (SHH1), which binds to H3K9me. The transcripts produced by Pol IV are then copied by RDR2, which interacts with the chromatin remodeler CLASSY1, into dsRNA that is diced into 24 nt siRNAs by DCL3. The siRNAs are protected from degradation through methylation by HEN1 and incorporated into AGO4.

In a second phase, Pol V mediates *de novo* DNA methylation of the target loci by transcribing a scaffold RNA with complementarity to the AGO4-bound siRNAs. ChIP-sequencing experiments showed that Pol V is targeted to transposons and

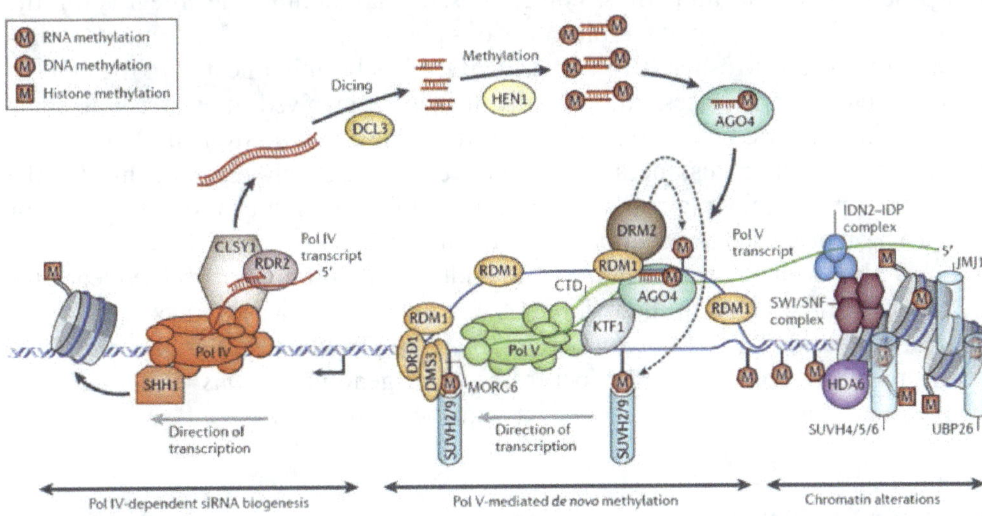

◘ **Fig. 6.6** Schematic representation of the canonical RdDM pathway involving the plant-specific RNA polymerases Pol IV and Pol V. The three phases of establishing silent chromatin at target loci, mostly transposable elements and repeats but also transgenes, are shown. In phase one, 24 nt siRNA are generated through a Pol IV-dependent process (left). In the second phase, the target loci get *de novo* methylated through a Pol V-mediated mechanism (middle) and in the last phase, heterochromatin is established through repressive histone modifications, nucleosome repositioning, and changes in higher order chromatin structure (right). For details, see main text. (From Matzke and Mosher 2014)

genomic repeats that are associated with DNA methylation and 24 nt siRNAs. However, about 25% of the Pol V loci do not show these features, indicating that Pol V occupancy is not sufficient for RdDM. Pol V recruitment to target loci may involve interactions with SU(VAR)3-9 HOMOLOG2 (SUVH2) and SUVH9, which both bind to methylated DNA. Pol V transcription is likely facilitated through the unwinding of DNA by the chromatin remodeler DEFECTIVE IN RNA-DIRECTED DNA METHYLATION1 (DRD1) and the single strand DNA-binding ability of RNA-DIRECTED DNA METHYLATION1 (RDM1) that, together with DEFECTIVE IN MERISTEM SILENCING3 (DMS3) and MICROCHIDIA6 (MORC6), may establish and maintain the unwound state. AGO4 is recruited to the locus through interactions with KOW DOMAIN-CONTAINING TRANSCRIPTION FACTOR1 (KTF1) and binding to Pol V itself. The RDM1 protein brings the DOMAINS REARRANGED METHYLTRANSFERASE2 (DRM2) together with AGO4, leading to *de novo* DNA methylation of regions with homology to the siRNAs.

Finally, in the third phase, a repressive chromatin state is established. This involves the repositioning of nucleosomes through a SWI/SNF chromatin remodeling complex (see book ▶ Chap. 2 of Paro) that interacts with a complex composed the Pol V scaffold RNA-binding proteins INVOLVED IN DE NOVO2 (IND2) and IND2 PARALOGUE (IDP). Active histone marks are removed by HISTONE DEACETYLASE6 (HDA6), histone demethylase JUMONJI14 (JMJ14), and UBIQUITIN-SPECIFIC PROTEASE26 (UNP26), followed by the deposition of repressive histone modifications, e.g., H3K9me, by SUVH4, SUVH5, and SUVH6. The silent state may be further reinforced through changes in higher order chromatin conformation mediated by the activity of MORC1 and MORC6.

Mutations affecting the RdDM machinery are viable and fertile and do typically not show gross developmental aberrations. However, the pathway is important for transposon repression and, thus, contributes to maintaining the integrity of the genome. Furthermore, it plays a role in various physiological and developmental processes (Matzke and Mosher 2014). Thus, RdDM has been implicated in the defense against bacteria and viruses. Additionally, it may be involved in establishing some gametic imprints (see book ▶ Chap. 5 of Grossniklaus). Moreover, siRNAs can be transported over short distances from cell to cell but also long-distance through the vasculature, potentially playing a role in intercellular communication. This has for instance been proposed to occur between cells in the female gametophyte and pollen, and may also occur between embryo and endosperm, although direct evidence for the transport of siRNAs in these tissues is missing. Finally, RdDM has been shown to occur in hybrids, where the production of siRNAs from one parental genome can affect previously non-methylated loci in the other genome in *trans*, either at allelic sites or at distant loci if there is sufficient sequence homology.

6.4 Paramutation

Paramutation is a classical epigenetic phenomenon that is now understood to involve RdDM. It was simultaneously discovered by Alexander Brink and Edward Coe in maize as well as Robert Hagemann in tomato (Brink 1956; Coe Jr 1959; Hagemann 1958). Paramutation describes a paradox of genetics where certain alleles do not show the expected segregation pattern based on the rules of Mendel. The discoverers found that a 'genetic' change or 'conversion' occurred in certain heterozygous combinations. For instance, Coe worked with alleles of the *booster1* (*b1*) locus responsible for anthocyanin pigmentation of the plant and reported "an effect on plant color in maize, interpreted as a regular conversion of one allele at the *B* locus by another [...]. The effect is a continuing one in that the 'converted' *B*, termed *B'*, is also regularly able to 'convert' newly introduced *B''*. When Coe crossed intensely pigmented, purple plants homozygous for the *booster1-Inten*se (*B-I*) allele to weakly pigmented, green plants homozygous for the *B* allele, all progeny was weakly pigmented. This could, in principle, be explained by dominance of *B* over *B-I*, but the progeny also remained weakly pigmented after repeated outcrossing to *B-I* plants. Thus, given there was no segregation, the *B-I* allele appeared to have changed its activity to that of the weakly pigmented *B* allele and was termed *B'* (◘ Fig. 6.7). The non-Mendelian segregation was confirmed in test-crosses of *B-I*/*B* heterozygotes to unpigmented plants homozygous for the loss-of-function *b* allele. Instead of the expected 1:1 segregation of intensely pigmented *B-I*/*b* and weakly pigmented *B*/*b* plants, all progeny was weakly pigmented, confirming the conversion of *B-I* to *B'*. Already in these early studies, Coe found that the *B-I* allele can spontaneously give rise to the *B'* allele and Brink found the altered state of the *R'* allele of the *coloured1* (*r1*) locus to revert in some of the progeny. Such instability or metastability is now considered a hallmark of an epigenetic state, which is much less stable than genetic mutations.

Nowadays, the conversion of one allele into another is understood as a change in the epigenetic control of the expression state of the locus and is called paramutation. Paramutation is defined as a directed, heritable change in gene expression that is invoked by interactions between specific alleles. Paramutation can occur between genetically identical or distinct alleles. The alleles that incite paramutation are called paramutagenic, those that are susceptible to it are paramutable, and those that do

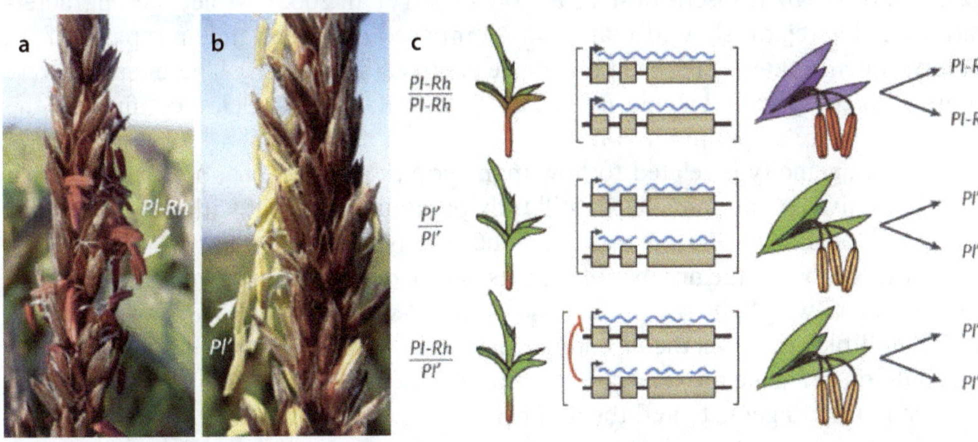

◘ **Fig. 6.7** Paramutation illustrated by the genetic behavior of the *purple plant1-Rhoades* (*Pl-Rh*) allele required for anthocyanin pigmentation of the plant. **a** Pigmented anthers of a *Pl-Rh* plant. **b** Unpigmented anthers of a *Pl'* plant. (**a**, **b** from Hollick 2017). **c** Schematic representation of paramutation at the *pl1* locus. Diploid genotypes and seedling coloration are shown on the left, anther color and haploid genotypes of the gametes produced are shown on the right. In the middle, transcription from the *pl1* locus is represented: the strongly expressed *Pl-Rh* allele, the weekly expressed *Pl'* allele with broken wavy lines representing unstable mRNAs, and the interaction between the paramutagenic *Pl'* and the paramutable *Pl-Rh* (red arrow), leading to reduced expression of *Pl-Rh* and conversion to *Pl'*. (From Hollick 2010)

not participate in paramutation are neutral. As Coe described, the converted paramutable allele then becomes paramutagenic itself and can incite secondary paramutation. This interesting epigenetic phenomenon has been described for several endogenous loci in tomato and maize, including *b1*, *r1*, *purple plant1* (*pl1*), and *pericarp color1* (*p1*), all which are transcriptional regulators of flavonoid biosynthesis. But there is also an increasing number of transgenes that show a paramutation-like behavior, both in plants and animals, including *D. melanogaster*, *C. elegans*, and mice (Hollick 2017).

6.4.1 The *cis*-Regulatory Elements Controlling Paramutation and *trans*-Acting Factors Link Paramutation to RdDM

Most paramutation systems in maize are metastable and the expression state of an allele can spontaneously change in both directions at variable frequencies. At the *b1* locus, however, *B′* alleles are formed spontaneously from *B-I*, while the *B′* state is extremely stable. In genetic screens metastability can be problematic as it may generate false positives. The stability of the *B′* state was one of the reasons to use this system in various genetic screens to identify *cis*-regulatory regions and *trans*-acting factors of paramutation.

For instance, the *cis*-regulatory regions required for high expression of *B-I* were localized by looking for recombinants where high expression was acquired by a neutral allele. Likewise, the *cis*-elements required for paramutation were identified by isolating recombinants where a neutral allele had become paramutagenic. Mapping of the recombination breakpoints showed that the enhancer for high *b1* expression and the region mediating paramutation coincided and were located about 100 kb upstream of the *b1* transcriptional start site (Stam et al. 2002). Molecular characterization of this region showed that it was composed of seven tandem repeats of an 853 bp sequence that is unique in the maize genome (◘ Fig. 6.8). These repeats were identical in the paramutagenic *B′* and the paramutable *B-I* alleles, confirming the epigenetic nature of paramutation.

Paramutagenicity is related to how many copies of the repeat are present in an allele. With five repeats, an allele is still fully paramutagenic but with three, paramutagenicity is reduced, and neutral alleles have only a single repeat (◘ Fig. 6.8). A relationship between the number of repeats and paramutagenicity was also observed at the *r1* locus, where the paramutagenic alleles *R-stippled* and *R-marbled* both consist of multiple copies of the *r1* coding region and flanking sequences, while alleles with only one *r1* gene are neutral. Deletion derivatives showed a direct correlation between paramutagenicity and the number of *r1* genes (Hövel et al. 2015).

Given that both *B-I* and *B′* have the same seven repeats, the difference between them must lie in their epigenetic state. Indeed, at several loci, the paramutable and paramutagenic alleles were shown to be differentially methylated (◘ Fig. 6.8). Importantly, upon paramutation, the paramutable allele gained DNA methylation, implying methylation is associated with and possibly even triggering the epigenetic switch underlying paramutation. Thus, during paramutation, the *cis*-regulatory regions mediate *trans*-homolog interactions between the two alleles, leading to a 'transfer' of DNA methylation. The tandem repeats of *B′* are not only hypermethyl-

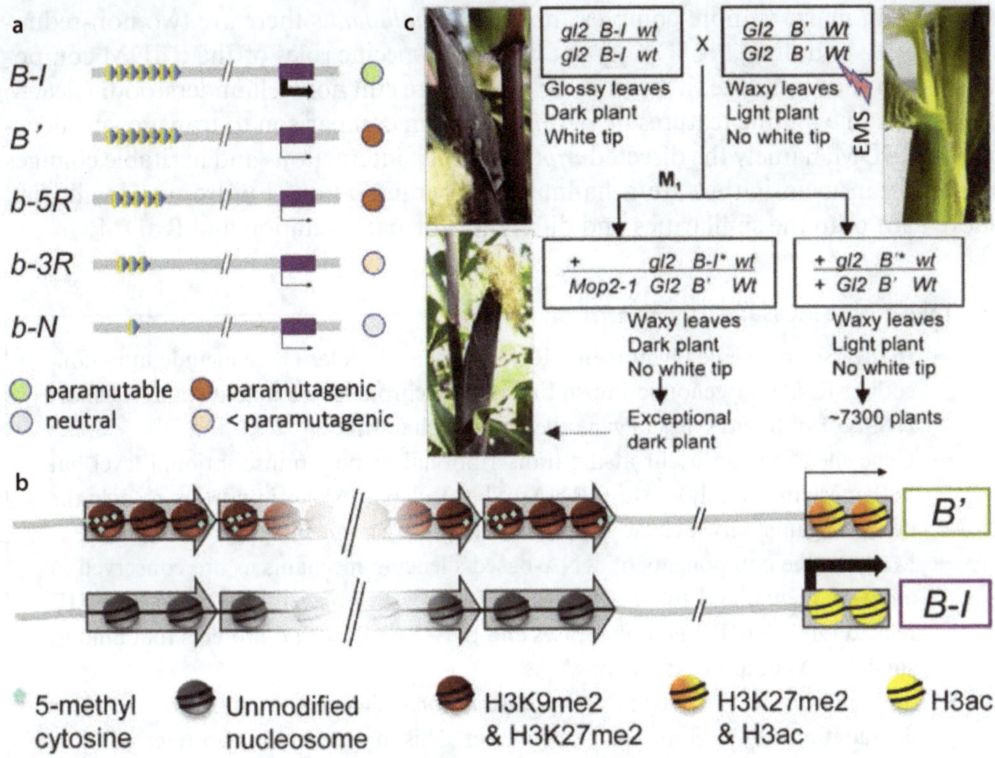

Fig. 6.8 Features of *cis*-elements required for paramutation at the *b1* locus and genetic screen for *trans*-acting factors. **a** Repeat region ~100 kb upstream of the *b1* coding sequence. The paramutagenic *B-I* and paramutable *B'* alleles have seven identical tandem repeats while the neutral *b-N* allele has only one. A reduction to five repeats maintains full paramutagenicity while, with three repeats, the allele is less paramutagenic. **b** Epigenetic features of *B'* and *B-I*, which show differential DNA methylation and carry distinct histone marks at the tandem repeats (only four repeats are show for simplicity) (**a**, **b** from Hövel et al. 2015). **c** Scheme of the genetic screen for *Mediator of paramutation* (*Mop*) mutants using EMS pollen mutagenesis. The two parental lines carry wild-type or mutant alleles of *glossy2* (*gl2*) and *white tip* (*wt*), respectively, that flank the *b1* locus. *B'** indicates a *B-I* allele that was paramutated to *B'* in wild-type plants, *B-I** indicates a *B-I* allele exposed to *B'* in the presence of *Mop2-1*, which prevents paramutation. (From Sidorenko et al. 2009)

ated as compared to those of *B-I* but they also carry the repressive histone marks H3K9me2 and H3K27me3, similar to silenced transposons (◘ Fig. 6.8).

Using the powerful genetic tools in maize, screens for second-site mutations in factors required for the establishment and/or maintenance of paramutation were conducted (Hollick 2017). Looking for mutants that abolish the reduced pigmentation levels observed in *Pl'/Pl'* and *B'/B'* plants, parallel screens for *required to maintain repression* (*rmr*) and *mediator of paramutation* (*mop*) mutants were performed (◘ Fig. 6.8). The molecular cloning of some of these mutants provided a clear link to gene silencing by RdDM. The *mop1* locus was found to encode a homolog of RDR2, which is required for the generation of Pol IV-dependent 24 nt siRNAs in *A. thaliana*. Moreover, the *rmr6/mop3* locus encodes the largest subunit of Pol IV, and *rmr7/mop2* the second largest subunit, which is part of both Pol IV and Pol V, fully establishing the link between paramutation and RdDM. Indeed, all known *mop* and *rdr* mutants do not only suppress the reduced pigmentation in *Pl'/Pl'* plants but also fail to accumulate the corresponding 24 nt siRNAs (Hollick 2017). However, the situation in maize is more complex than in *A. thaliana* as there are two non-redundant Pol IV and three Pol V subtypes. Thus, the specific roles of the RdDM components identified in maize through these screens are still not well understood. Clearly, there are some unique features of paramutation in comparison to transposon silencing by RdDM, namely the directed *trans*-homolog interactions and heritable changes occurring in meiosis that are a hallmark of paramutation. Future work will shed more light onto the similarities and differences of paramutation and RdDM.

Take-Home Message

- In diverse epigenetic phenomena, RNAs play a key role. These include long non-coding RNAs in genomic imprinting and X chromosome inactivation, or short RNAs of different types in gene silencing mechanisms.
- Gene silencing can occur at the transcriptional or post-transcriptional level but both mechanisms rely on short RNAs. The mechanisms play important roles in the defense against viruses as well as in maintaining genome integrity.
- Many of the components of RNA-based silencing mechanisms are conserved in animals, plants, and fungi. These include some key players like the RNase III-related DICERLIKE endonucleases and the ARGONAUTE proteins that bind to small RNAs and slice target mRNAs.
- Small RNAs are involved in post-transcriptional silencing, targeting mRNAs for degradation or inhibiting their translation. This mechanism is also referred to as RNA interference (RNAi) and has been instrumental for large-scale screens in cultured cells both in fundamental and applied research.
- Transcriptional gene silencing, for instance to repress transposable elements and repeats, is also based on small RNAs – siRNAs or piRNAs – across a wide variety of organisms. In plants, this is achieved by RNA-dependent DNA-methylation, a complex pathway involving the plant-specific RNA polymerases Pol IV and Pol V.
- Paramutation involves *trans*-homolog interactions that cause a heritable repression of the paramutable allele directed by the paramutagenic allele. Similar effects were observed in hybrids where bringing together divergent (epi)genomes can cause DNA methylation changes in *trans*.
- Paramutation relies on *cis*-regulatory sequences that can be far away from the coding sequence and consist of repeats. The epigenetic state of these repeats differs between paramutable and paramutagenic alleles and is established and/or maintained by the RNA-dependent DNA methylation pathway.

References

Bernstein E, Caudy AA, Hammond SM, Hannon GJ (2001) Role for a bidentate ribonuclease in the initiation step of RNA interference. Nature 409:363–366. https://doi.org/10.1038/35053110.

Bestor TH, Chandler VL, Feinberg AP (1994) Epigenetic effects in eukaryotic gene expression. Genesis 15:458–462. https://doi.org/10.1002/dvg.1020150603

Borges F, Martienssen R (2015) The expanding world of small RNAs in plants. Nat Rev Mol Cell Biol 16:727–741. https://doi.org/10.1038/nrm4085

Brink RA (1956) A genetic change associated with the *R* locus in maize which is directed and potentially reversible. Genetics 41:872–889. PMID: 17247669

Coe EH Jr (1959) A regular and continuing conversion-type phenomenon at the *B* locus in maize. Proc Natl Acad Sci USA 45:828–832. https://doi.org/10.1073/pnas.45.6.828

Ernst C, Odom DT, Kutter C (2017) The emergence of piRNAs against transposon invasion to preserve mammalian genome integrity. Nat Commun 8:1411. https://doi.org/10.1038/s41467-017-01049-7

Fire A, Xu S, Montgomery MK, Kostas SA, Driver SE, Mello CC (1998) Potent and specific genetic interference by double-stranded RNA in *Caenorhabditis elegans*. Nature 391:806–811. https://doi.org/10.1038/35888

Gonsalves D, Gonsalves C, Ferreira S, Pitz K, Fitch M, Manshardt R, Slightom J (2004) Transgenic virus resistant papaya: from hope to reality for controlling papaya ringspot virus in Hawaii. APSnet Feature Story July 2004. https://pdfs.semanticscholar.org/6db3/b62d6f8348fc780e4b5a3a8f-293da309e52c.pdf?_ga=2.17733369.1394956602.1601814746-498759226.1601380287

Gottula J, Fuchs M (2009) Toward a quarter century of pathogen-derived resistance and practical approaches to plant virus disease control. Adv Virus Res 75:161–183. https://doi.org/10.1016/S0065-3527(09)07505-8.

Guo S, Kemphues KJ (1995) *par-1*, a gene required for establishing polarity in *C. elegans* embryos, encodes a putative Ser/Thr kinase that is asymmetrically distributed. Cell 81:611–620. https://doi.org/10.1016/0092-8674(95)90082-9

Hagemann R (1958) Somatische Konversion bei *Lycopersicon esculentum* Mill. Z Vererbungslehre 89:587–613. PMID: 1360483

Hamilton AJ, Baulcombe DC (1999) A species of small antisense RNA in posttranscriptional gene silencing in plants. Science 286:950–952. https://doi.org/10.1126/science.286.5441.950.

Hokaiwado N, Takeshita F, Banas A, Ochiya T (2008) RNAi-based drug discovery and its application to therapeutics. IDrugs 11:274–278. PMID: 18379962.

Hollick JB (2010) Paramutation and development. Annu Rev Cell Dev Biol 26:557–579. https://doi.org/10.1146/annurev.cellbio.042308.113400

Hollick J (2017) Paramutation and related phenomena in diverse species. Nat Rev Genet 18:5–23. https://doi.org/10.1038/nrg.2016.115

Hövel I, Pearson NA, Stam M (2015) *Cis*-acting determinants of paramutation. Sem Cell Dev Biol 44:22–32. https://doi.org/10.1016/j.semcdb.2015.08.012

Hutvagner G, Simard MJ (2008) Argonaute proteins: key players in RNA silencing. Nat Rev Mol Cell Biol 9:22–32. https://doi.org/10.1038/nrm2321.

Jiang L, Wei C, Li Y (2012) Viral suppression of RNA silencing. Sci China Life Sci 55:109–118. https://doi.org/10.1007/s11427-012-4279-x

Johanson TM, Lew AM, Chong MMW (2013) MicroRNA-independent roles of the RNase III enzymes Drosha and Dicer. Open Biol 3:130144. https://doi.org/10.1098/rsob.130144

Martienssen R, Moazed D (2015) RNAi and heterochromatin assembly. Cold Spring Harb Perspect Biol 7:a019323. https://doi.org/10.1101/cshperspect.a019323

Matzke MA, Mosher RA (2014) RNA-directed DNA methylation: an epigenetic pathway of increasing complexity. Nat Rev Genet 15:394–408. https://doi.org/10.1038/nrg3683

Matzke MA, Primig M, Trnovsky J, Matzke AJM (1989) Reversible methylation and inactivation of marker genes in sequentially transformed tobacco plants. EMBO J 8:643–649. PMID: 16453872

Mette MF, Aufsatz W, van der Winden J, Matzke MA, Matzke AJ (2000) Transcriptional silencing and promoter methylation triggered by double-stranded RNA. EMBO J 19:5194–5201. https://doi.org/10.1093/emboj/19.19.5194

Napoli C, Lemieux C, Jorgensen R (1990) Introduction of a chimeric chalcone synthase gene into petunia results in reversible co-suppression of homologous genes *in trans*. Plant Cell 2:279–289. https://doi.org/10.2307/3869076

Pikaard CS, Mittelsten Scheid O (2014) Epigenetic regulation in plants. Cold Spring Harb Perspect Biol 6:a019315. https://doi.org/10.1101/cshperspect.a019315

Romano N, Macino G (1992) Quelling: transient inactivation of gene expression in *Neurospora crassa* by transformation with homologous sequences. Mol Microbiol 6:3343–3353. https://doi.org/10.1111/j.1365-2958.1992.tb02202.x

Sen GL, Blau HM (2006) A brief history of RNAi: the silence of the genes. FASEB J 20:1293–1299. https://doi.org/10.1096/fj.06-6014rev

Sidorenko L, Dorweiler JE, Cigan AM, Arteaga-Vazquez M, Vyas M et al (2009) A dominant mutation in *mediator of paramutation2*, one of three second-largest subunits of a plant-specific RNA polymerase, disrupts multiple siRNA silencing processes. PLoS Genet 5(11):e1000725. https://doi.org/10.1371/journal.pgen.1000725

Song JJ, Smith SK, Hannon GJ, Joshua-Tor L (2004) Crystal structure of Argonaute and its implications for RISC slicer activity. Science 305:1434–1437. https://doi.org/10.1126/science.1102514.

Stam M, Mol JNM, Kooter JM (1997) The silence of genes in transgenic plants. Ann Bot 79:3–12. https://doi.org/10.1006/anbo.1996.0295

Stam M, Belele C, Dorweiler JE, Chandler VL (2002) Differential chromatin structure within a tandem array 100 kb upstream of the maize *b1* locus is associated with paramutation. Genes 16:1906–1918. https://doi.org/10.1101/gad.1006702

Wassenegger M, Heimes S, Riedel L, Sänger HL (1994) RNA-directed de novo methylation of genomic sequences in plants. Cell 76:567–576. https://doi.org/10.1016/0092-8674(94)90119-8

Wei K, Wu L, Chen Y, Wang Y, Liu Y, Xie D (2013) Argonatue protein as a linker to command center of physiological processes. Chin J Cancer Res 25:430–441. https://doi.org/10.3978/j.issn.1000-9604.2013.08.13

7

Regenerative Medicine: Cell Reprogramming and Reconditioning

Contents

What You Will Learn in This Chapter

During regenerative processes, cells are required to restructure parts of a damaged or worn-out organ and tissue. Here, you will become acquainted with the strategies that organisms developed to provide the material for tissue and organ repair. On the one hand, somatic cells can become dedifferentiated to increase their developmental potential and produce the plasticity required to replace the entire cellular complexity of a damaged part. On the other hand, organisms retain organ-specific stem cells with a restricted developmental potency and use these to provide the "spare parts" for replacing damaged cells. In all cases, a substantial reprogramming of the epigenome of these cells accompanies the restructuring process. *In vitro* strategies have been developed to drive cells back to a pluripotent state, allowing a better understanding of the underlying chromatin adjustments and providing a rich source for cellular therapies.

7.1 Types of Regenerative Phenomena

Epigenetic mechanisms mostly sustain long-term processes, like maintenance of gene expression patterns, keeping either the maternal or the paternal X chromosome repressed within a cell lineage, maintaining cellular memory over developmental time, or tightly silence transposon-riddled genome sections during the life of a cell and of its progenitors. Many of these mechanisms have evolved to be exceptionally stable, approaching sometimes the perseverance of the genetic system. Yet, organisms also adapt to changing environments and physiological requirements and, therefore, have to adjust or reverse some of these long-term epigenetic commitments. A prime example requiring restructuring of the epigenomic landscape is the process of regeneration. Damaged body parts, organs, or tissues can be faithfully repaired contributing to the survival of the adult organism. In adults, development has been completed and cell types and the anatomy of tissues in the organs is maintained with minimal changes. Additionally, the process of ageing leads to a degradation of tissue function. Hence, new strategies to provide the materials necessary for the repair process evolved. The body can carry a reservoir of stem cells from which replacement of parts can form through cell differentiation, by re-running developmental cascades to achieve the necessary structures and functions. Alternatively, differentiated cells in the neighborhood can change their own identity and attain the morphological structures and physiological functions required to replace the damaged parts. In low complexity organs, hyperproliferation of the basic structural and functional constituents can often fill the gaps left by the damage. This chapter deals with the obstacles that regenerative processes encounter when epigenetically encoded, stable states have to be changed and reprogrammed.

While plants have extraordinary regenerative capacities and there is a large body of literature describing the underlying epigenetic processes (Ikeuchi et al. 2019), this chapter will focus on animal regenerative mechanisms (Tanaka and Reddien 2011; Carlson 2005). Different phenomena have been observed in animal phyla, like morphallaxis in Planaria or Hydra, approaching the repair capacities observed in the plant world. Tissue fragments retain the ability to reconstitute an entire functional organism or organ. In general, compensatory growth restores the size of an organ by increasing cell size (compensatory hypertrophy) or by accelerating cell division (compensatory hyperplasia). Limited to the usage of the existing pool of differentiated cells, such regulatory processes can be observed after damage of a number of human organs, like liver or kidney. Physiological regeneration describes the continuing replacement of damaged or ageing cells or body parts. In humans, examples for this form of tissue homeostasis are the process of blood cell replacement, the renewal of the lining of the intestine, or the surface of the skin.

7.1.1 Regenerating from a Blastema

Reparative regeneration represents the most spectacular form of restoration observed in higher animals. The capability to regrow an entire limb structure after amputation, as observed in newts, is certainly such a remarkable process, regarded with envy by us humans. Not surprisingly, these reparative processes were studied for decades by scientists in a number of model organisms, like the amphibian axolotl[1] or newts. The repair and regenerative process after amputation of the limb can be described in the following series of key events (Carlson 2005):

1. Wound Healing: Epidermal migration to cover the amputation surface and prevent infections by pathogens.
2. Demolition and Phagocytosis: Inflammatory response and enzymatic removal of matrix components just beneath the wound epithelium.
3. Dedifferentiation: The loss of differentiated tissue types beneath the wound epithelium and the appearance of embryonic-looking (dedifferentiated) cells in the same area.
4. Blastema Formation: The aggregation of the dedifferentiated cells into a structure reminiscent of embryonic limb structure.
5. Morphogenesis and Growth: Growth of the blastema and shaping and differentiation of the blastemal cells into the cells and tissues comprising the normal limb. In larger appendages, after morphogenesis is completed, growth of the miniature regenerate continues until it is of the same size as the original limb structure that was lost.

Central to the process is the formation of a blastema containing the progenitor cells for the regrowth of the full appendage (epimorphic regeneration). Cells forming the blastema derive from tissues surrounding the wound. Though resembling the morphology of embryonic cells, these precursors nevertheless appear not be fully dedifferentiated to a pluripotent state (see below) but gain a multipotent condition required to rebuilt the resident organ or tissue. Lineage tracing experiments in axolotl reveal that the cells of the blastema retain certain developmental restrictions and only have a limited differentiation potential (Tanaka and Reddien 2011). By tracing cells from each major tissue during axolotl limb regeneration, dermal cells turned out to have the potency to form cartilages and tendons, but never switched the fate across embryonic

1 Interestingly, axolotl does loose regenerative ability when it is hormonally forced to develop to the adult form. This is an interesting case how regenerative ability has been apparently selected along with a developmental arrest. This might add support to a view of a re-running of developmental programs to a restricted extent for regenerative purposes.

germ layers. Other blastema cells derived from epidermal cells, muscles, Schwann cells, and cartilage cells, retained their original tissue identity during regeneration. Hence, a rather limited dedifferentiation and change in fate appears to occur at the blastema.

7.1.2 Changing Potency by Transdifferentiation

Other processes of regeneration, omitting the formation of a blastema but still utilizing partial dedifferentiation steps, have been observed. For example, zebrafish hearts can regenerate after amputation of up to 20% of the ventricle. During the process, the newly formed heart muscle cells are derived from the dedifferentiation and proliferation of pre-existing cardiomyocytes. Dedifferentiation promotes proliferation and the activation of embryonic cardiogenesis genes. Conversely, a decrease in expression of sarcomeric components leads to the disassembly of sarcomeric structures (Barrero and Izpisua Belmonte 2011).

In other cases, a complete change in cellular fate has been described and termed *"trans-differentiation"*. Glucagon-producing α-cells can transdifferentiate into pancreatic insulin-producing β-cells after ablation of β-cells with diphtheria toxin. This transition process seems to be direct, not involving undifferentiated intermediates[2]. Transdifferentiation can also occur through an undifferentiated intermediate, however, as observed during the regeneration of the eye lens in newts. Lens regeneration takes place through transdifferentiation of the pigmented epithelial cells (PECs) of the iris. After removal of the lens, the PECs re-enter the cell cycle, dedifferentiate, and lose their characteristic pigmentation. During this early stage, the expression of developmental regulators, such as the transcription factors Pax6 and Sox2, is activated. Later, the dedifferentiated proliferating PECs start to express crystallins and differentiate into lens fibers.

Taken together, these results suggest that the process of dedifferentiation and transdifferentiation during regeneration involves the silencing of tissue-specific genes, as well as the induction of genes involved in embryonic programs and the control of the cell cycle. These changes in gene expression might facilitate the acquisition of a limited extended plasticity that allows cells to proliferate and rearrange into the new structures by a limited iteration of developmental pathways. Potentially, embryonic genes involved in regeneration might remain poised for activation in regenerating animals, but irreversibly silenced in non-regenerating animals. The questions remaining are: what signaling mechanisms drive this change in potency and what mechanisms repattern the cells to reproduce a full organ?

7.1.3 Signaling in the Blastema

Despite substantial interest in understanding the basic mechanisms regulating regenerative capacity, in many of the classical model organisms it has been difficult to study the underlying molecular mechanisms. Conversely, *Drosophila melanogaster* offers substantial advantages for regenerative studies since it is less complex than amphibians and

2 Also during normal development α and β-cells are derived from an initially bipotential progenitor. The potential of the progenitor for making α is switched later in development to β-cells. Reflecting again a copy mechanism of developmental steps in regeneration.

humans and a wide palette of research tools is available. *Drosophila*, unlike some other arthropods, is not able to regenerate damaged legs or wings in adults. However, the larvae harbour regeneration abilities in the imaginal discs (see also book ► Chap. 3 of Paro). Imaginal discs can regenerate to form normal adult appendages even after massive lesion of disc cells are caused by X-rays irradiation at larval stages or by local induction of cell death (Hariharan and Serras 2017; Ahmed-de-Prado and Baonza 2018).

Regeneration of the imaginal discs is also observed under *ex vivo* conditions (◘ Fig. 7.1). When imaginal discs are manually fragmented, transplanted, and cultured in the abdomen of an adult fly, the disc cells at the wound site regenerate the missing parts. Remember that the imaginal disc cells consist of already specifically determined but undifferentiated cells until metamorphosis, hence they are not uniform cells. Each imaginal disc established during embryogenesis is destined to follow a specific developmental pathway. The regional identities and specific cell fates in each disc are precisely determined in a stepwise manner throughout the larval stage.

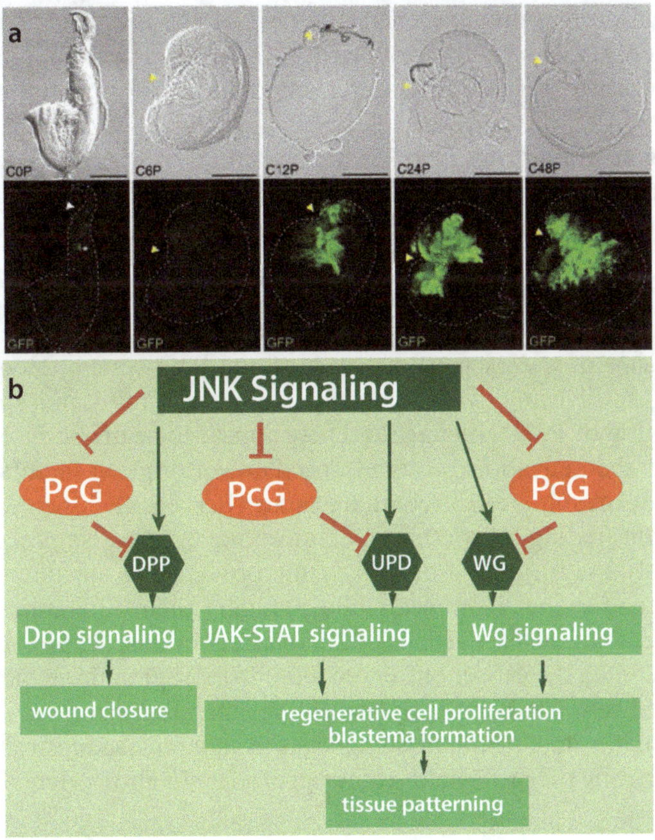

◘ Fig. 7.1 Signaling in the blastema of the regenerating imaginal disc. **a** Leg imaginal disc dissected from third instar larva at 100 h after egg deposition, wounded and cultured in adult abdomen (ex vivo). Activation of JNK signaling in cells shown by a continuous labeling of a GFP-reporter (bottom panel) in the blastema at different time points after wounding; (C0P) 0 hours (un-cultured), (C6P) 6 hours, (C12P) 12 hours, (C24P) 24 hours, and (C48P) 48 hours of culturing. Yellow arrowheads indicate the fragmented positions. Scale bar = 100 µm. **b** JNK signaling at the wound site is required to downregulate PcG silencing and up-regulate ligands for Dpp, JAK-STAT, and Wg signaling cascades. Dpp signaling initiates wound closure. The concerted action of JAK-STAT and Wg signaling induces cell proliferation and tissue repatterning in the blastema

At the time of damage, the expression of master regulatory genes shows already an intricate developmental pattern. The steps in the regeneration process of fragmented imaginal discs are, in principle, analogous to those of amphibian limb regeneration, consisting of wound healing (closure), localized cell proliferation (regeneration blastema formation), and pattern formation. Collectively, the observations demonstrate that disc regeneration induces limited cellular reprogramming, enabling the reconstitution of the lost tissue while disc identity is maintained independently of the local environment.

Studies of imaginal disc transplantation have uncovered the process of transdetermination, i.e., neighbouring groups of cells in regeneration blastema sometimes become more plastic and acquire alternative organ identities from different imaginal discs (comparable to transdifferentiation; see book ► Chap. 3 of Paro). This suggested that the mechanism of cellular memory has to be changed and reprogrammed during the regenerative process. Indeed, disc regeneration and transdetermination are coupled to the regulation of Polycomb (PcG) function. Down-regulation of PcG function, as monitored by the derepression of a silent PcG regulated reporter gene, was observed in proliferating, regenerating cells of the blastema (Lee et al. 2005). Furthermore, only cells with compromised PcG function were found to transdetermine more frequently, suggesting that PcG modulation is a prerequisite for cellular reprogramming.

The healing at wound sites involves the activation of the Jun N-terminal kinase (JNK) signalling pathway in several rows of cells at the edge of the wound (◘ Fig. 7.1). The blastema cells are derived from cells in which JNK has been activated. It was found that the JNK signalling pathway directly controls the down-regulation of PcG silencing (Lee et al. 2005). Clonal activation of the JNK pathway in imaginal discs was sufficient to reduce PcG-mediated silencing function, as can be visualized by ectopic expression of a *Hox* gene, a target of the PcG system. Hence, the down-regulation of PcG silencing by JNK signaling appears to result in a specific, yet, not global reactivation of PcG target genes. These appear to be driven into a poised state and only upon the activation by tissue-specific transcription factors, the cells are directed towards the required differentiated state (Beira et al. 2018).

An important finding shows that in epimorphic regenerative processes observed in many animals, the formation of a blastema provides the starting source of cells used for the repatterning. However, while changing their differentiation state to a more multipotent character, the cells of the blastema never acquire a fully pluripotent capacity. Unlike the exogenous expression of reprogramming factors to induce pluripotent stem cells (see below), the natural regeneration signals are able to reprogram the chromatin state of the blastema cells to more lineage-restricted progenitor states, circumventing the problem of tumorigenic deregulation often observed in pluripotent cells.

7.2 Stem Cells in the Adult

A different concept for providing cells to regenerating parts are stem cells. By definition, the division of a stem cell will produce another stem cell as well as a precursor for a particular lineage. Hence, on the one side, the constant availability of a stem cell pool is guaranteed. On the other side, the stem cell has to maintain the developmental potential to produce all the cell types of the lineage(s) that is contributing to. In mammals at the early blastocyst stage, cells from epiblast of the inner cell mass can be maintained in culture. They have the capacity to develop and contribute to cell types of all three germ layers when subjected *in vitro* to a differentiation program or transplanted back into the corresponding developmental stage of the organism. As such, they are pluripotent since they have the potential to develop into any part of the organism, except the extraembryonic tissue giving rise to the placenta. Conversely, adult organisms maintain pools of stem cells with a more restricted developmental capacity (multipotent). Here, the classical example are hematopoietic stem cells. The "fluid" nature of the blood system allowed the study of the various cell types and their pedigrees in great detail. The original source is the multipotential hematopoietic stem cells in the bone marrow. The fact that we can use normal hematopoietic stem cells to repopulate the full blood system in a leukemic patient crippled with cancerous blood cells is a clear indication that our body carries a pool of cells with highly specific regenerative capacities. Is this true for every organ? The search for an answer to this question has occupied generations of scientists. Hampered by their apparently limited cell number and the difficulty to isolate and cultivate them, for long time it has been challenging to identify molecular markers characteristic for adult stem cells. In particular, by analysing organs with a high cellular turnover like skin or gut, caused, for example, by mechanical or chemical strain, the corresponding source of adult stem cells could be localized and the differentiation path their progeny follows could be studied. The intestinal epithelium is composed at the basis of crypts encompassing the cradle of stem cells (expressing, among others, the marker Lgr5) (Tetteh et al. 2015). Progenitors of the stem cells leaving the cradle are subjected to intricate signalling cascades and start to differentiate along their path to the villus (the intestinal surface exposed to the inside of the stomach). Lgr5-expressing cells can be cultivated *in vitro*. Eventually, they will self-assemble into a mini-gut organoid, manifesting their developmental capability to structure a specific part of an organ. This example demonstrates that also the intestine harbours its own "spare parts" to counter the constant degradation and turnover of cells and thereby maintains tissue homeostasis. Other pools of adult stem cells have been identified in different organs (Clevers and Watt 2018), but their cultivation *in vitro* and molecular characterization has been exceedingly difficult. This is in part caused by the exclusive neighbourhood required by the stem cells. Stem cells clearly must have a different epigenetic status than most of their differentiated neighbours. They seem to only be able to maintain this status in a special environment, termed "niche". A niche represents a highly protective microenvironment, located in a specific anatomic location where the stem cells are found. At these sites adult stem cells are maintained in a quiescent state by a combination of cell-cell or cell-matrix interactions, endocrine, paracrine, and autocrine signalling pathways, specific physical interactions, neuronal connections and particular physico-chemical conditions. However, after tissue injury or to maintain tissue homeostasis, the surrounding micro-environment actively signals to stem cells to induce either self-renewal or differentiation to form cells required for regeneration. As soon as stem cells leave the niche, they are subjected to differentiation signals, become first transient amplifying cells, and then start to develop fates similar to their neighbours. Recreating niche conditions in a dish to cultivate the multipotent stem cells has been a major challenge, though.

7.3 Sources of Pluripotent Stem Cells

Why bother with finicky adult stem cells if pluripotent embryonic stem cells can produce all the cell types of a mammalian organism? Indeed, already 40 years ago, mouse embryonic stem cells (mESC) from the inner cell mass of blastocysts could be established as an *in vitro* culture. Eventually, these cells could be induced *in vitro* by specific combinations of developmental factors to produce almost all cell types of a mouse, demonstrating their pluripotent capacity. Experiments with human embryonic stem cells demonstrated a similar capacity. The possibility to cultivate and drive theses pluripotent cells through specific differentiation paths to produce functionally and morphologically elaborate cell types, offered seemingly extraordinary opportunities for regenerative medicine. In parallel, alternative methods to collect or produce pluripotent cells were developed (◘ Fig. 7.2). Besides the described cell mass from blastocysts before implantation, *in vivo*, two other sources of mouse pluripotent stem cells exist. Embryonic germ cells can be derived from primordial germ cells during mid-gestation (mouse embryonic day 8.5–12.5). Germline-derived pluripotent stem cells can be isolated from spermatogonial stem cells of neonatal and adult testes[3].

The study of murine germ cell development allowed to observe the dramatic reshuffling of epigenetic information not only during germ cell maturation but also after fertilization (◘ Fig. 7.3) (Saitou et al. 2012). The massive changes of the epigenetic landscape detected over time reflect the substantial remodelling germ cells undertake to bring the epigenome back to the ground state (see also book ► Chap. 3 of Paro).

In primordial germ cells, different waves of DNA de- and re-methylation together with chromatin modifying mechanisms ensure that genomic imprints are reset, inactive X chromosomes reactivated, and somatic epigenetic signatures eliminated (see also book ► Chap. 5 of Grossniklaus). Similar mechanisms will apply to the somatic cell lineages. Hence, organisms have developed very elaborate mechanisms to reprogram, in a substantial and directional manner, the epigenetic landscape of some of their cell types.

In the past decades, three *in vitro* methods have been developed to generate cells with a pluripotent capacity (Gaspar-Maia et al. 2011). Fusions between a differentiated somatic cell and an embryonic stem cell reprograms the nucleus restricted in developmental potential to a pluripotent state. Here, reprogramming processes could be studied *in vitro* and the ensuing changes of the epigenetic states could be pursued at the molecular level. The disadvantage, in particular for therapeutic purposes, was that the resulting cell contains both genomes of the fusion partners and, thus, a tetraploid state is obtained. Transferring a nucleus from a somatic cell to an enucleated oocyte also changes the epigenome to a pluripotent ground state. Somatic cell nuclear transfer was most spectacularly demonstrated with the birth of the cloned sheep Dolly, providing evidence that a transferred somatic cell nucleus from the mammary

---

3 Primordial germ cells (PGCs) are specified from the epiblast and form the cell lineage that will develop into the gametes. EG cells are derived from PGCs before sex specific differentiation by inductive signals form the forming gonad from mesodermal genital ridges, which forms the somatic cells of testis and ovary. Spermatogonial stem cells are precursors in the male germ line and give rise to the spermatocytes.

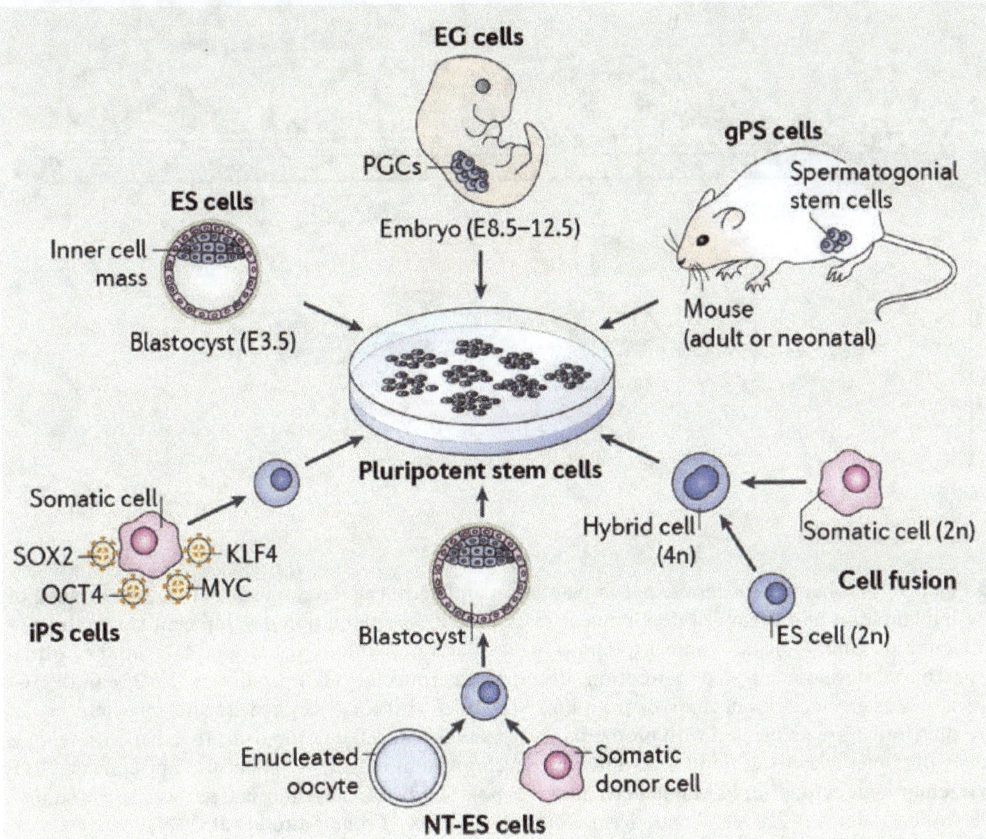

Fig. 7.2 Pluripotent stem cells can be derived from several sources. There are three sources of pluripotent stem cells *in vivo* (see the figure, top half). Embryonic stem (ES) cells are derived from the inner cell mass of the blastocyst, before embryo implantation. Embryonic germ (EG) cells are derived from primordial germ cells (PGCs) during mid-gestation (embryonic days 8.5–12.5 in the mouse) and germline-derived pluripotent stem (gPS) cells are derived from spermatogonial stem cells of neonatal and adult testes. In addition, three major routes for somatic cell reprogramming to pluripotency have been described (see the figure, bottom half): fusion between a somatic cell and an ES cell giving rise to reprogrammed hybrid cells; the generation of nuclear transfer embryonic stem (NT-ES) cells, produced by reprogramming of a somatic nucleus by an enucleated oocyte, which is then cultured to the blastocyst stage to allow derivation of ES cells; and the production of induced pluripotent stem (iPS) cells, derived by somatic cell overexpression of reprogramming transcription factors, most commonly OCT4 (also known as POU5F1), Sry-box containing gene 2 (SOX2), myelocytomatosis oncogene (MYC) and Krüppel-like factor 4 (KLF4). (From (Gaspar-Maia et al. 2011)

epithelium of an ewe could be fully reprogrammed to restart development and produce an entire mammal. Reprogramming utilizing transcription factors is nowadays the most powerful technique to obtain large quantities of cells with a pluripotent developmental potential, however. Furthermore, this method is ethically also less controversial because it does not require access to mammalian oocytes.

Developmental biology studies identified many transcription factors that act as master regulators for defining cell fates (Graf and Enver 2009). Forced expression of these master regulators can override determined cell identities and respecify cells to a new cell type. An example is Myf5, whose expression in fibroblasts will induce the formation of contractile muscle cells. In parallel with the respecification, reprogram-

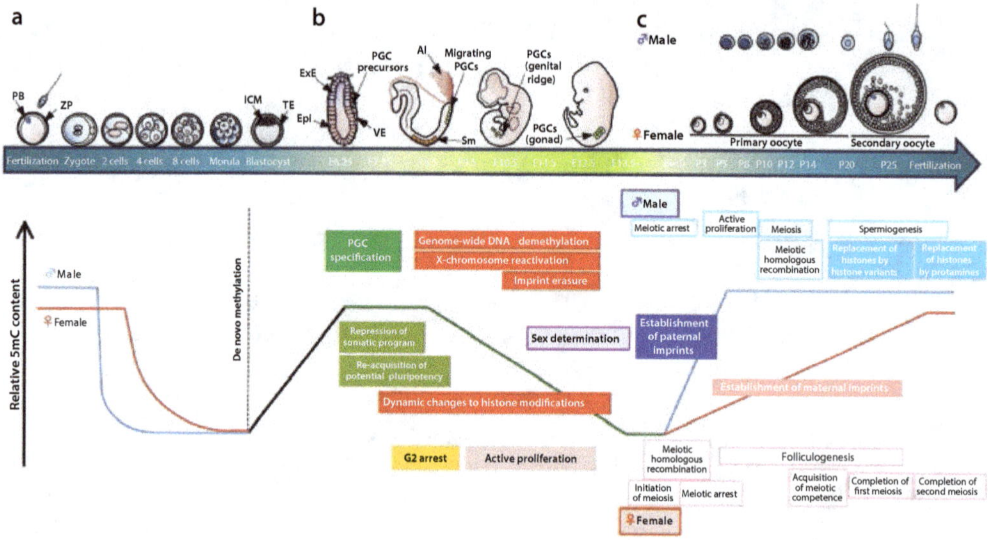

Fig. 7.3 A schematic of mouse pre-implantation and germ cell development. (Top) A schematic of pre-implantation and germ-cell development in mice. **a** Pre-implantation development stages; **b** post-implantation embryonic development, following blastocyst implantation at around E4.5; and **c** postnatal germ cell development and maturation. Primordial germ cell (PGC) precursors (E6.25) and PGCs are shown as green circles in embryos from E6.25 to E12.5. (Bottom) Key genetic and epigenetic events are shown that are associated with pre-implantation and germ cell development, together with relative levels of 5-methylcytosine (5mC) at different developmental stages. Al allantois, Epi epiblast, ExE extraembryonic ectoderm, ICM inner cell mass, PB polar body, PGCs primordial germ cells, Sm somite, TE trophectoderm, VE visceral endoderm, ZP zona pelucida. (From (Saitou et al. 2012)

ming of the cell's epigenome can be observed using biochemical tools. Based on these same principles, Yamanaka and co-workers were able to change a fully differentiated mouse fibroblast back to a pluripotent stem cell (Takahashi and Yamanaka 2006). Mouse embryonic stem cells had been extensively characterized and their transcriptome and underlying regulatory network carefully and comprehensively catalogued. Using this information, they selected a combination of four transcription factors which, when expressed artificially in fibroblasts, reversed the entire differentiation path back to the early pluripotent state. However, this process was very inefficient (less than 1% of the cells became pluripotent, termed *induced pluripotent stem cells (iPSC)*), already providing a hint that differentiated cellular states are very stable and, only under special circumstances and on rare occasions, revert to another state. The trick was to equip the fibroblast with a fluorescent reporter that was controlled by the *Nanog* promoter (*Nanog* + green fluorescent protein (GFP)) which would only become activated when the cell entered the pluripotent state. Subsequently, iPS cells were also derived from human fibroblasts. In the following years, this technique became standard in many laboratories, allowing the production of large quantities of iPS cells from different sources, thereby opening outstanding perspectives for regenerative medicine (see below). Interestingly, over these many years of research, the low efficiency of iPS cells reprogramming could not be substantially improved. This illustrates that the cellular memory encoded in the somatic state of the cell is remarkably stable and highly resistant to induced reprogramming (see book ▶ Chap. 3 of Paro). Hence, the question is: what are the underlying epigenetic hurdles that need to be removed or remodelled for cell reprogramming?

7.4 Chromatin Dynamics During Reprogramming

Already at the microscopic level, the nuclei of ESC show a different morphology compared to nuclei from differentiated cells. In particular, heterochromatic regions are much more discernible and compact in somatic cells, indicating that one accompanying force of differentiation is the compaction of specific regions of chromatin. This suggests that reprogramming of fates need not only changes local chromatin landscapes to activate or repress different sets of genes, but also requires large scale reshuffling of chromatin domains and chromosomal territories. An example of this is the reactivation of the entire repressed X chromosome in female mammalian cells[4] (see book ► Chap. 4 of Wutz). The availability of a technique inducing the reprogramming of somatic chromatin to a pluripotent state, hence, offered a great opportunity to study the underlying processes in substantial detail. The facility of growing cells in culture enabled biochemical studies, using classical transcriptome analysis methods, ChIP, and other chromatin mapping techniques to elucidate changes in the epigenetic landscape during the reprogramming process. The observed complexity of cellular transitions during iPSC induction provide a glimpse of why reprogramming takes a sizeable amount of time and why only few cells manage to reach the pluripotent state. The entire process appears to be stochastic and the individual steps difficult to predict. It is thought that the four transcription factors [Oct4, Sox2, KLF4, and Myc (OSKM)] induce a cascade of changes in the transcription program of the starting cell population. Subsequently, the readjustment of gene expression profiles with the accompanying changes in chromatin and DNA modifications cause: (i) inhibition of somatic regulators, (ii) induction of cellular proliferation, (iii) inhibition of senescence and apoptosis pathways, (iv) activation of pluripotency loci, (v) acquisition of factor independence, (vi) immortalization and finally reprogramming of telomere structures, X chromosome reactivation, and cellular memory erasure (see ◘ Fig. 7.4). The exogenously introduced transcription factors are capable of establishing a self-regulatory loop, eventually activating the corresponding endogenous loci and thereby sustaining a self-propagating stem cell identity. This is a crucial step as exogenously expressed factors become often silenced over time. Additionally, in order to differentiate, iPS cells, as for ESC, need to downregulate pluripotency factors and this can be done only when the endogenous genes are expressed.

OSKM acting as pioneer factors in the reprogramming process require the support from different epigenetic modifiers including histone post-translational modifying enzymes, nucleosome remodeling factors, histone chaperones, and DNA modifying enzymes (◘ Fig. 7.4).

Initially, OSKM bind to active somatic genes, eventually inducing the change of their active histone modifications like H3K4me1 and H3K27ac into repressive marks. Conversely, genes characteristic for embryonic stages lose their negative histone marks and, subsequently, become reactivated. During the entire process, a massive replacement of histone variants can be observed, either by reactivating the

4 Interestingly, reactivation of the inactive X chromosome is observed in mouse but not human iPS cells. This suggests that the developmental state of ESC/iPSC from humans is likely very different from that of mouse and corresponds to a more determined cell state with partially established chromatin patterns.

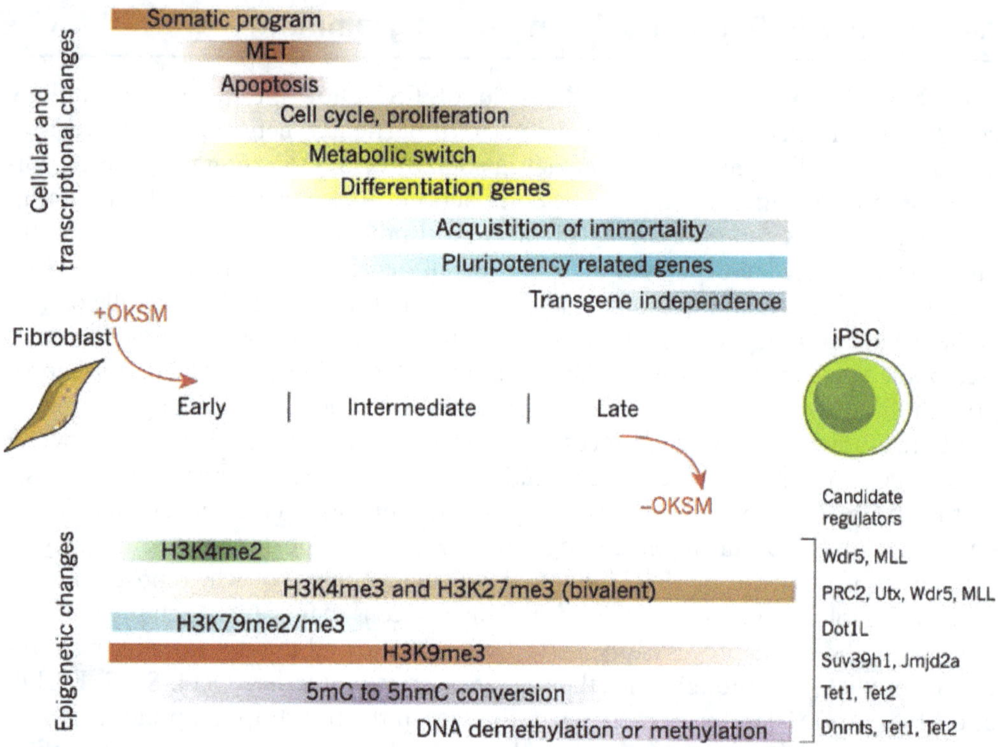

Fig. 7.4 Dynamics of key molecular events during direct reprogramming. A summary of cellular, transcriptional and epigenetic changes (colored bars) that occur during induced pluripotent stem cell (iPSC) formation from fibroblasts and examples of candidate regulators that have been associated with the depicted chromatin marks in the context of direct reprogramming (right). Red arrows indicate the time points of exogenous factor induction (+OKSM) and withdrawal (–OKSM). 5hmC 5-hydroxymethylcytosine, 5mC 5-methylcytosine, MET mesenchymal-to-epithelial- transition, OKSM Oct4, Klf4, Sox2, c-Myc. (From (Apostolou and Hochedlinger 2013))

quiescent part of the genome (i.e. H2A.X, TH2A/2B) or by removing repressing variants like macroH2A (see book ► Chap. 1 of Wutz and book ► Chap. 2 of Paro). Similarly, histone chaperones play an import role in the genome-wide realignment of histones and nucleosomes. Inhibition of CAF-1 can, for example, increase the reprogramming efficiency. Differentiation is inherently connected to a specific deposition of CpG dinucleotide methylation (see book ► Chap. 1 of Wutz). The entire DNA methylation and demethylation apparatus is heavily engaged over the length of the reprogramming process. DNA dioxygenases like TET1/2, which are involved in the demethylation process, can partly replace OSKM in chromatin remodeling. Not surprisingly, modulating the amount and activity of chromatin factors (writer, readers, and erasers) has a substantial impact on the efficiency of the reprogramming process. For example, adding the DNA methylation inhibitor 5-aza-cytidine during reprogramming substantially increases the production of iPS cells more than 30-fold. As expected, the reprogramming process does not only lead to a readjustment of the somatic epigenetic landscape to an ESC-like state but also to a complete re-organization of the 3D chromatin architecture and topology (Apostolou and Stadtfeld 2018).

7.5 Regenerative Therapies

Stem cell technologies certainly represent a major breakthrough development in regenerative medicine (Cherry and Daley 2012). The availability of large quantities of cells with a pluripotent developmental capacity provide exceptional opportunities to establish novel cell therapies. Additionally, iPS cells can be derived from patients with specific syndromes and used as models for studying disease. In this way, specific drug screens and diagnostic tests for personalized medicine can be developed. A very exciting development based on stem cells in general and iPS cells in particular is the possibility to create organoid structures *in vitro* (Akkerman and Defize 2017). Originally, small three-dimensional (3D) tissue structures were produced *in vitro* using tissue culture cells. In most cases these cells are considered to have a tumorigenic state as they can proliferate indefinitely in the dish and therefore the usefulness of these mini-organs was limited. The availability of self-renewing embryonic stem cells and eventually of iPS cells allowed the development of complex culturing and differentiation conditions, leading to organoid structures of remarkable morphological and functional resemblance to the real organ. Here also, the availability of iPS cells from patients as a source for specific organoid construction allows researchers to study the effect of human syndromes on the early development of the corresponding organ.

Despite the extraordinary enthusiasm accompanying iPSC technology, only few clinical trials have so far tested the medical usefulness of stem cell transplants. As mentioned, normal regenerative processes in almost all complex animals do not involve cells with a pluripotent developmental potential but organ- and tissue-specific multipotent stem and progenitor cells. For good reasons, organisms restrain and keep under strict control cells in a pluripotent state (see germ cell development). Cancer cells often attain and maintain their highly proliferative and destructive capability by dedifferentiation, thereby increasing their epigenetic resemblance to embryonic stem cells (see book ► Chap. 3 of Paro and ► Chap. 8 of Santoro). Incidentally, the first pluripotent cells to be cultivated *in vitro* were from teratocarcinomas (cancers of the germ line). Hence, there is only a fine line between normal development from a pluripotent state and degenerated development. Indeed, when transplanted into the skin of mice, iPS cells develop into carcinomas, but injection into blastocysts allows them to contribute to normal development. Additionally, it was found that iPS cells do not always erase all epigenetic marks from the somatic state from which they were derived, leaving, for example, distinctive somatic DNA methylation patterns. This imposes on the starting cells an unwanted epigenetic history and may, in certain conditions, lead to aberrant differentiation behaviour, aging onset, or susceptibility to degeneration. Using iPSC derived from the same person to install regenerative processes potentially circumvents the well-known problems of organ rejection. However, the delivery and integration of iPS cells to damaged parts in the organism is still a major challenge.

Second generation iPSC technology is attempting to reduce some of these problems by inducing the reprogramming process *in vivo* (Martinez-Redondo and Izpisua Belmonte 2020). CRISPR/Cas9 technology produced forms of the Cas9 enzyme which allows the alteration of histone modifications and other epigenetic marks in a site-specific manner. Potentially, this approach could aid the entire reprogramming process to the pluripotent state but also partial reprogramming to intermediate or alternative differentiation states. After all, most known regenerative processes in complex organisms use the latter for repopulating damaged organs or tissues, either by using highly specialized stem cells with a restricted potential or by reassigning necessary new fates to cells at the injury site. Thus, regenerative biology and medicine will need to understand these processes in much better detail in order to be able to use cell therapies in a controllable and predictive fashion.

It is very clear that the epigenetic landscape of cells used in therapies is of crucial importance. Hence, an epigenetic profiling of stem cells used in clinical context appears as an obligatory requirement as is the sequencing of the genome to eliminate cell lineages with potential deleterious mutations. Indeed, technically we are approaching capabilities to comprehensively screen cells for genetic and epigenetic integrity. What we are still lacking is a better knowledge of how natural regenerative processes are able to reprogram the epigenome of somatic cells to the desired organ-specific multipotent state. This would permit to apply cellular therapies in a much more controlled fashion.

Take-Home Message

- Amphibians and many other animals can regenerate complex body structures through a blastema (epimorphic regeneration). In the blastema, ordered signaling events reprogram cells for patterning and regrowth processes.
- In regenerating *Drosophila* imaginal discs, JNK is activated at the wound site and in the blastema, resulting in downregulation of PcG silencing and, as a consequence, re-activation of many silenced PcG target genes, like the homeotic master regulators.
- Dedifferentiation and transdifferentiation of cells/tissues is also utilized in vertebrates to produce cells with structures and functions necessary to regenerate a wound or an organ damage in their neighborhood.
- Organ-specific adult stem cells contribute to physiological regeneration required for tissue and organs, like gut or skin, subjected to constant mechanical and chemical stress. They undergo specific differentiation programs and extended epigenetic reprogramming to maintain tissue homeostasis.
- Hematopoietic stem cells are used in the clinic to treat leukemia patients. Stem cells from other organs are difficult to isolate and cultivate, requiring specific conditions found in stem cell niches.
- Pluripotent stem cells can be induced to differentiate into any other cell type. iPSC technology reprograms differentiated cells to induced pluripotent stem cells. Transcription factor-driven reprogramming encompasses numerous pathways requiring reversal of many epigenetic processes.
- Some of the inefficiency of iPSC reprogramming can be explained by the need to change the cellular memory system of differentiated cells to the embryonic state. A highly coordinated and sequential erasure of chromatin marks and DNA methylation patterns must occur during reprogramming.
- iPS cells can be used for cell therapies. Patient-derived iPS cells can be used to study disease or for the development of new drugs. A promising new path is the engineering of self-organizing organoid structures using iPS cell cultures.

References

Ahmed-de-Prado S, Baonza A (2018) Drosophila as a model system to study cell signaling in organ regeneration. Biomed Res Int 2018:7359267. https://doi.org/10.1155/2018/7359267

Akkerman N, Defize LHK (2017) Dawn of the organoid era: 3D tissue and organ cultures revolutionize the study of development, disease, and regeneration. BioEssays: News Rev Mol Cell Develop Biol 39(4):1600244. https://doi.org/10.1002/bies.201600244

Apostolou E, Hochedlinger K (2013) Chromatin dynamics during cellular reprogramming. Nature 502:462

Apostolou E, Stadtfeld M (2018) Cellular trajectories and molecular mechanisms of iPSC reprogramming. Curr Opin Genet Dev 52:77–85. https://doi.org/10.1016/j.gde.2018.06.002

Barrero MJ, Izpisua Belmonte JC (2011) Regenerating the epigenome. EMBO Rep 12(3):208–215. https://doi.org/10.1038/embor.2011.10

Beira JV, Torres J, Paro R (2018) Signalling crosstalk during early tumorigenesis in the absence of Polycomb silencing. PLoS Genet 14(1):e1007187. https://doi.org/10.1371/journal.pgen.1007187

Carlson BM (2005) Some principles of regeneration in mammalian systems. Anat Rec B New Anat 287(1):4–13. https://doi.org/10.1002/ar.b.20079

Cherry ABC, Daley GQ (2012) Reprogramming cellular identity for regenerative medicine. Cell 148(6):1110–1122. https://doi.org/10.1016/j.cell.2012.02.031

Clevers H, Watt FM (2018) Defining adult stem cells by function, not by phenotype. Ann Rev Biochem 87(1):annurev-biochem-062917-012341. https://doi.org/10.1146/annurev-biochem-062917-012341

Gaspar-Maia A, Alajem A, Meshorer E, Ramalho-Santos M (2011) Open chromatin in pluripotency and reprogramming. Nat Publ Group 12(1):36–47. https://doi.org/10.1038/nrm3036

Graf T, Enver T (2009) Forcing cells to change lineages. Nature 462(7273):587–594. https://doi.org/10.1038/nature08533

Hariharan IK, Serras F (2017) Imaginal disc regeneration takes flight. Curr Opin Cell Biol 48:10–16. https://doi.org/10.1016/j.ceb.2017.03.005

Ikeuchi M, Favero DS, Sakamoto Y, Iwase A, Coleman D, Rymen B, Sugimoto K (2019) Molecular mechanisms of plant regeneration. Annu Rev Plant Biol 70:377–406. https://doi.org/10.1146/annurev-arplant-050718-100434

Lee N, Maurange C, Ringrose L, Paro R (2005) Suppression of PcG by JNK signalling induces transdetermination in Drosophila imaginal discs. Nature 438:234–237

Martinez-Redondo P, Izpisua Belmonte JC (2020) Tailored chromatin modulation to promote tissue regeneration. Semin Cell Dev Biol 97:3–15. https://doi.org/10.1016/j.semcdb.2019.04.015

Saitou M, Kagiwada S, Kurimoto K (2012) Epigenetic reprogramming in mouse pre-implantation development and primordial germ cells. Development 139(1):15–31. https://doi.org/10.1242/dev.050849

Takahashi K, Yamanaka S (2006) Induction of pluripotent stem cells from mouse embryonic and adult fibroblast cultures by defined factors. Cell 126(4):663–676. https://doi.org/10.1016/j.cell.2006.07.024

Tanaka EM, Reddien PW (2011) The cellular basis for animal regeneration. Dev Cell 21(1):172–185. https://doi.org/10.1016/j.devcel.2011.06.016

Tetteh PW, Farin HF, Clevers H (2015) Plasticity within stem cell hierarchies in mammalian epithelia. Trends Cell Biol 25. Elsevier Ltd

8

Epigenetic Mechanisms in Cancer

Contents

What You Will Learn in This Chapter

Alterations in chromatin function and epigenetic mechanisms are a hallmark of cancer. The disruption of epigenetic processes has been linked to altered gene expression and to cancer initiation and progression. Recent cancer genome sequencing projects revealed that numerous epigenetic regulators are frequently mutated in various cancers. This information has not only started to be utilized as prognostic and predictive markers to guide treatment decisions but also provided important information for the understanding of the molecular mechanisms of epigenetic regulation in both physiological and pathological conditions. Furthermore, the reversible nature of epigenetic aberrations has led to the emergence of the promising field of epigenetic therapy that has already provided new therapeutic options for patients with malignancies characterized by epigenetic alterations, laying the basis for new and personalized medicine.

8.1 Epigenetics and Cancer

Cancer is a group of more than 100 different and distinct diseases in which abnormal cells divide without control and can invade nearby tissues. Cancer is the second leading cause of death globally. In the recent years, there have been tremendous efforts and remarkable advances to understand and treat this disease. In particular, the newest sequencing technologies have made possible to obtain a complete DNA sequence of large numbers of cancer genomes. These analyses identified genetic alterations not only between normal and cancer genomes but also between genomes of tumors from patients affected by the same type of cancer. This tremendous effort had provided relevant information for the history of tumor development at molecular level and led the basis for a personalized treatment approach that allows doctors to select treatments based on a molecular understanding of their disease.

The earliest indications of an epigenetic link to cancer originated from gene expression and DNA methylation studies. The past decade has seen a remarkable acceleration in the validation of the concept that cancer is not only a genetic disease but also one of epigenetic abnormalities. This view has been significantly strengthened by whole-genome sequencing results showing that numerous epigenetic regulators are frequently the target of mutations and epimutations in cancer cells, with an intriguing interplay between the two. The abundance of cancer mutations involving these genes practically affects all levels of epigenetic regulation, including key players in DNA methylation, histone modifications, and chromatin organization but also substrates for these modifications, such as histones. The identification of mutations in writers, readers, and erasers, as well as alterations in the epigenetic landscape in cancers, do not only imply a causative role for these factors in cancer initiation and progression but also provide potential targets for therapeutic intervention.

In this chapter, we describe the best studied epigenetic and chromatin alterations found in cancers (DNA methylation, histone H3K27 trimethylation, histone acetylation, and chromatin remodeling factors), how these alterations occur, and how they contribute to cancer initiation and progression. We also provide examples of the development of epigenetic inhibitors and strategies for their therapeutic application.

8.2 DNA Methylation and Cancer

Aberrant DNA methylation was the first epigenetic abnormality to be identified in human cancers (Feinberg and Vogelstein 1983). DNA methylation provides a stable gene silencing mechanism that plays an important role in regulating gene expression and chromatin architecture in association with histone modifications and other chromatin associated proteins (see book ► Chap. 1 of Wutz). Alterations detected in cancers can be promoter hypermethylation and the consequent silencing of tumor suppressor genes (► Sect. 8.2.1), global hypomethylation that has been associated with genomic instability (► Sect. 8.2.2), and alterations of DNA methylation at imprinting control regions (see book ► Chap. 5 of Grossniklaus) with a consequent loss of imprinting (► Sect. 8.2.3). Regions of low-density methylation near CpG islands, known as 'shores', exhibit great variation in DNA methylation, including hypomethylation and hypermethylation, across different types of cancers.

8.2.1 DNA Hypermethylation in Cancer

Many genes associated with CpG islands undergo *de novo* methylation in cancer. The "CpG island methylator phenotype (CIMP)" is defined as frequent methylation of multiple CpG islands and has been found in many types of cancers (◘ Fig. 8.1). The CIMP status is uniquely associated with specific clinicopathological characteristics in individual cancer types and has the potential to provide information for cancer diagnosis and the stratification of patients for specific therapeutic treatments.

Since DNA methylation is associated with gene repression, it has been suggested that new/aberrant acquisition of DNA methylation contributes to the repression of genes that are active in normal tissue and probably involved in tumor suppression

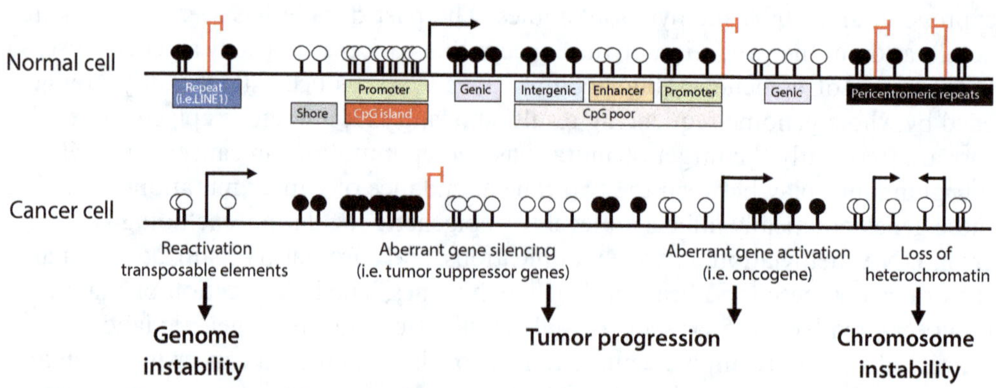

◘ **Fig. 8.1** Normal and cancer genomes exhibit distinct DNA methylation profiles. Schematic showing the difference in DNA methylation at defined gene regulatory elements and repeats between normal and cancer cells. These alterations include hypermethylation of CpG islands and hypomethylation at transposable elements and pericentromeric heterochromatin. These changes cause transcriptional silencing of tumor suppressor genes, the increase in the expression of oncogenes, and genome and chromosome instability. White circle, unmethylated CpG; black circle, methylated CpG

activities[1]. Accordingly, hypermethylation in cancers has been associated with the suppression of genes involved in cancer-related pathways, such as cell cycle control, DNA repair, apoptosis, and angiogenesis. The profiles of hypermethylation of the CpG islands in tumor suppressor genes are specific to the cancer type. Examples of this are the methylation at the promoter of the tumor suppressor gene *VHL* (associated with von Hippel-Lindau disease) in renal cancer, the cell cycle control gene *p16* in many types of cancer, the DNA mismatch repair MutL Homolog 1 (*MLH1*) gene in colorectal cancer, and breast-cancer susceptibility 1 (*BRCA1*) gene in breast and ovarian cancers. Although these are important examples of DNA methylation-mediated gene repression, it was also shown that a large fraction of genes acquiring DNA methylation in cancer were already transcriptionally repressed in the normal tissue. It was proposed that the repression of these genes was mainly acting through PcG proteins that establish H3K27me3 (see book ▶ Chap. 3 of Paro). It was also shown that genes targeted by DNA methylation in cancer have a low, poised transcription state in embryonic stem cells and are implicated in the maintenance of stemness and self-renewal in normal stem cells. Their reactivation occurs at defined stages of development through the removal of the repressive PcG-complexes. Thus, the acquisition of DNA methylation at genes that are already in a repressed state might serve to either further downregulate and/or establish a stable repression that prevents activation, thereby contributing to a stem cell-like state of cancer. However, mechanisms that target *de novo* methylation at genes critical to cancer remain yet elusive.

8.2.2 DNA Hypomethylation in Cancer

Loss of 5-methylcytosine (5mC) was the first epigenetic abnormality to be identified in human cancer. The loss of DNA methylation is mainly occurring at repetitive DNA sequences, coding regions, and introns (◘ Fig. 8.1). The extensive demethylation of DNA observed in tumor progression was suggested to be a source of the continually generated cellular diversity associated with cancers. During the development of a neoplasm[2], the degree of hypomethylation of genomic DNA increases as the lesion progresses from a benign proliferation of cells (hyperplasia)[3] to an invasive and potentially metastatic cancer[4] (Fraga et al. 2004).

1 Tumor-suppressor activities are mediated by tumor-suppressor genes, class of genes that regulate cell proliferation and their inactivation can play a profound role in the development of malignancies.

2 Neoplasm is an abnormal mass of tissue that forms when cells grow and divide more than they should or do not die when they should. Neoplasms may be benign (not cancer) or malignant (cancer). Benign neoplasms may grow large but do not spread into, or invade, nearby tissues or other parts of the body. Malignant neoplasms can spread into, or invade, nearby tissues. They can also spread to other parts of the body through the blood and lymph systems.

3 Hyperplasia refers to the increase in the number of cells in an organ or tissue due to loss of proliferation control. These cells appear normal under a microscope. They are not cancer but may become cancer.

4 *Metastatic* cancer is a cancer that has spread from the part of the body where it started (the primary site, primary cancer) to other parts of the body.

Hypomethylation of DNA in cancer has several implications, including gene activation, loss of imprinting, reactivation of transposable elements, and genome instability. DNA methylation is particularly concentrated at repetitive elements and may limit their genomic activity. Loss of DNA methylation can favor mitotic recombination, leading to deletions, translocations, and chromosomal rearrangements. Furthermore, hypomethylation of DNA in malignant cells can reactivate transposable elements, such as the long interspersed nuclear element LINE 1, which accounts for 17% of the human genome (◘ Fig. 8.1). The large number of these repeats can, when reactivated, promote translocations to other genomic regions, thereby affecting genome integrity. Hypomethylation can also occur at pericentromeric regions of the chromosome (near the site of attachment of the mitotic spindle) and rearrangements in the pericentromeric heterochromatin of chromosomes 1 or 16 are found in many types of cancers.

8.2.3 Loss of Imprinting Through Alterations of DNA Methylation

The Beckwith-Wiedemann syndrome (BWS) is an important example of disruption of genome imprinting through alteration in DNA methylation (see book ▶ Chap. 5 of Grossniklaus). BWS is characterized by macrosomia, macroglossia, and abdominal wall defects, and exhibits a predisposition to tumorigenesis (Soejima and Higashimoto 2013). In particular, the development of embryonal tumors (i.e. Wilms' tumor, hepatoblastoma, and rhabdomyosarcoma) is a recurring feature of BWS. The relevant imprinted chromosomal region in BWS that is epigenetically altered is 11p15.5, which consists of two imprinting domains, *CDKN1C/KCNQ1OT1* and *IGF2/H19* (◘ Fig. 8.2) (see also book ▶ Chap. 5 of Grossniklaus).

In the imprinting domain *CDKN1C/KCNQ1OT1*, *CDKN1C* (Cyclin Dependent Kinase Inhibitor 1C) is maternally expressed, whereas the long non-coding RNA *KCNQ1OT1* is paternally expressed. The imprinting control region (ICR) of *CDKN1C/KCNQ1OT1* domain is the differentially methylated region (DMR) KvDMR1, which overlaps with the promoter of *KCNQ1OT1*. KvDMR1 is methylated on the maternal allele and unmethylated on the paternal allele. Fifty percent of BWS patients display a loss of DNA methylation on KvDMR1 at the maternal allele, which leads to the expression of *KCNQ1OT1* and repression of the cell cycle inhibitor *CDKN1C*. Representative phenotypes of this epigenetic alteration include omphalocele and hemihyperplasia[5]. Of note is that downregulation of *CDKN1C* has been reported in several other cancers. It has been proposed that the repression of *CDKN1C* on the paternal chromosome is mediated by the long non-coding RNA (lncRNA) *KCNQ1OT1*, which coats the surrounding locus. *KCNQ1OT1*, through the interaction with H3K9- and H3K27-specific histone methyltransferases G9a and EZH1/2, the latter part of Polycomb Repressive Complex 2 (PRC2) (see book ▶ Chap. 3 of Paro), leads to the repression of *CDKN1C* in *cis*.

5 Omphalocele is a birth defect in which an infant's intestine or other abdominal organs are outside of the body because of a hole in the belly button area. The intestines are covered only by a thin layer of tissue. Hemihyperplesia refers to asymmetric body overgrowth of one or more body parts.

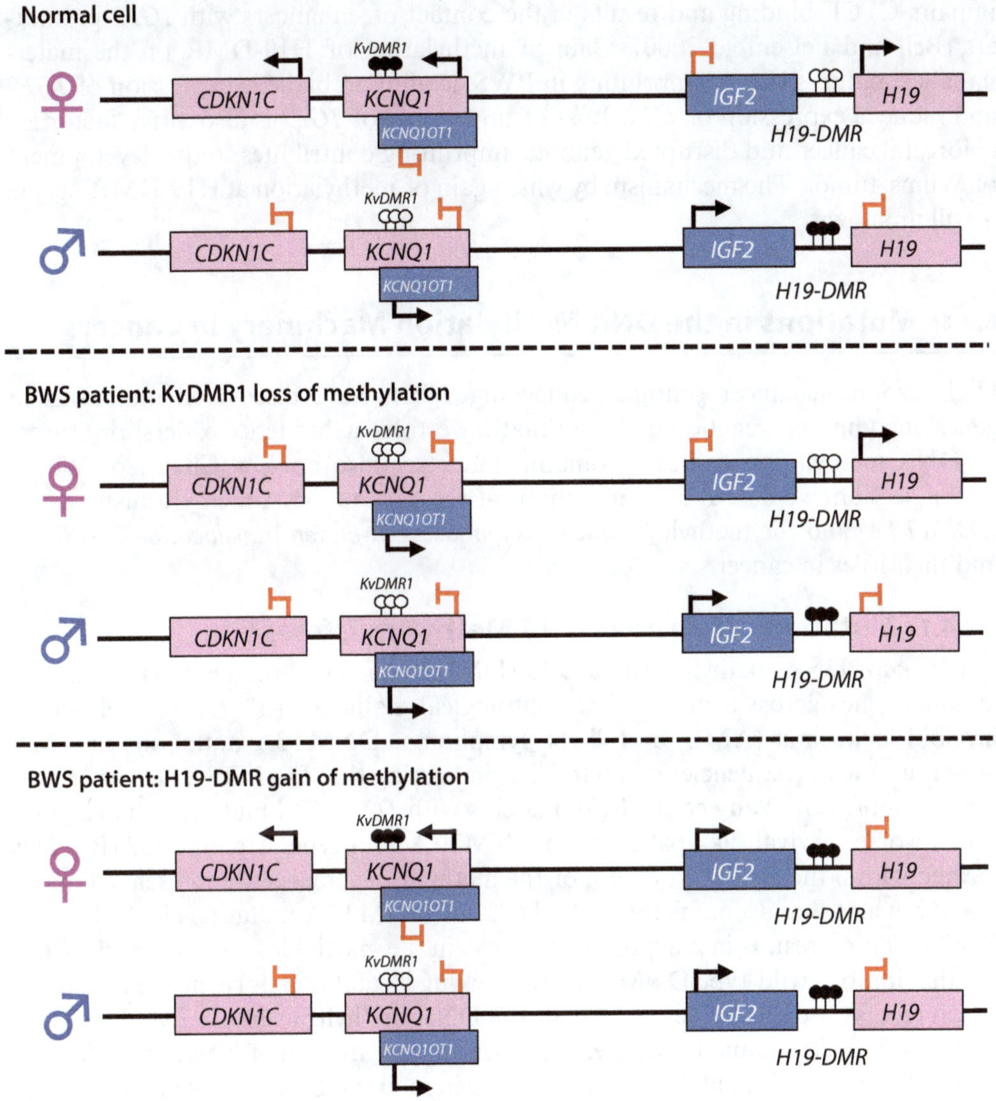

Fig. 8.2 Loss of imprinting through alteration of DNA methylation in cancer. Schema representing aberrant imprinting patterns at the chromosome 11p15 imprinting cluster found in Beckwith-Wiedemann Syndrome (BWS) patients. Loss of methylation at the differentially methylated regions (DMRs) KvDMR1 or H19-DMR and the resulting changes in the expression of imprinted genes are shown. Pink boxes represent maternally genes. Blue boxes are paternally expressed genes. White circle, unmethylated CpG; black circle, methylated CpG

In about 5% of BWS patients, gain of DNA methylation occurs on the normally unmethylated maternal H19-DMR, which is the ICR of *IGF2*/*H19* domain (Fig. 8.2). H19-DMR is located 2 kb upstream of *H19*. In normal cells, differential parental methylation of H19-DMR leads to paternal expression of *IGF2* (the insulin-like growth factor gene), and maternal expression of the non-coding RNA gene *H19*. On the maternal allele, *IGF2* is repressed due to CTCF binding the unmethylated H19DMR. The insulator formed by CTCF blocks enhancers downstream of *H19* from accessing *IGF2* promoters (see also ► Chap. 5 of Grossniklaus). On the paternal allele, *IGF2* is expressed since H19-DMR is methylated, which impairs CTCF binding and results in the contact of enhancers with *IGF2* promoters (Bell and Felsenfeld 2000). Gain of methylation of H19-DMR on the maternal allele induces loss of imprinting in BWS, leading to biallelic expression of *IGF2* and reduced expression of *H19*. Loss of imprinting of *IGF2* is also a risk factor for colorectal cancer and disrupted genomic imprinting contributes to the development of Wilms' tumor. The mechanism by which gain of methylation at H19-DMR occurs is still unknown.

8.2.4 Mutations in the DNA Methylation Machinery in Cancers

High-resolution cancer genome sequencing efforts have discovered mutations in genes encoding epigenetic regulators that have roles as writers, readers, or erasers of DNA methylation and/or chromatin states (see also book ▶ Chap.1 of Wutz). In this section, we describe the mutations at the *de novo* DNA methyltransferase 3a (*DNMT3A*) and the methylcytosine dioxygenase *Ten-eleven translocation 2* (*TET2*) and their roles in cancer.

8.2.4.1 Mutations of *de novo* DNA Methyltransferase 3a

The *de novo* DNA methyltransferase 3a (DNMT3A) is one of the most frequently mutated genes across a range of haematological malignancies[6], especially in acute myeloid leukaemia (AML) and T cell lymphoma. DNMT3A mutations in AML were found with frequencies of up to 22%, indicating that DNMT3A acts as critical tumor suppressor (Yang et al. 2015). Patients with *DNMT3A* mutations had significantly worse survival than patient with DNMT3A wild-type. Arginine 882 (R882) is located within the catalytic domain of the methyltransferase and represents a mutational hotspot that accounts for around 60% of DNMT3A mutations in AML. The R882 mutation results in a hypomorphic protein (i.e. partial loss of enzymatic activity) that inhibits wild-type DNMT3A by blocking its ability to form active tetramers, which represent the most active form of the DNA methyltransferase.

DNMT3A was found to be essential for the differentiation of hematopoietic stem cells (HSCs). Conditional deletion of *Dnmt3a* in mouse HSCs promoted self-renewal over differentiation and HSCs with a complete knock-out of the *Dnmt3a* gene (*Dnmt3a*-KO) dramatically outcompete their wild-type counterparts and accumulate in the bone marrow (◘ Fig. 8.3). *Dnmt3a*-KO HSCs display significant genome-wide hypomethylation. In particular, HSC-associated enhancers display dominant hypomethylation in *Dnmt3a*-KO HSCs. Accordingly, *Dnmt3a*-KO HSCs show upregulation of HSC multipotency genes and downregulation of differentiation factors, and their progeny exhibit global DNA hypomethylation and incomplete repression of HSC-specific genes.

It has been proposed that *DNMT3A* mutations serve as a pre-leukaemic lesion since somatic *DNMT3A* mutations arise in HSCs many years before malignancies develop. It has also been suggested that the enhanced self-renewal capacity of HSCs with *DNMT3A* mutations would be advantageous for second pro-oncogenic hits that drive tumor development (Brunetti et al. 2017). *DNM3A* mutations were

6 *Hematologic malignancies* are classified according to which type of blood cell is affected.

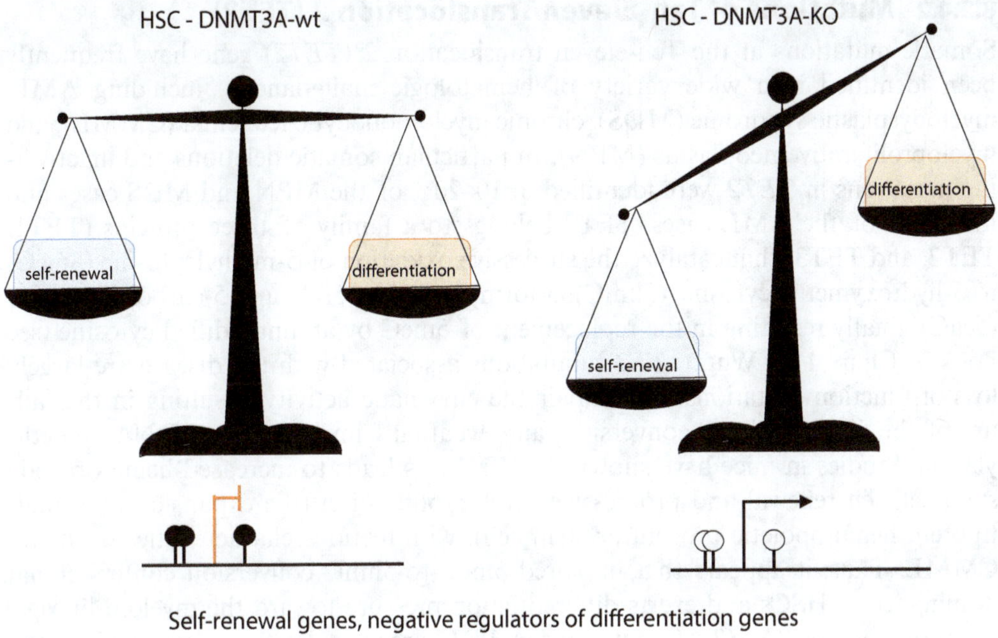

Fig. 8.3 DNMT3A controls hematopoietic stem cell fate. DNMT3A is a critical factor for the epigenetic silencing of hematopoietic stem cell (HSC) regulatory genes. *Dnmt3a*-KO HSCs show substantial CpG island hypermethylation, upregulate HSC multipotency genes, and downregulate differentiation factors. DNMT3A loss progressively impairs HSC differentiation and expands HSC numbers in the bone marrow

found to negatively correlate with other known AML alterations, such as chromosomal translocations t(15;17), t(8;21) and inversion inv(16), rearrangements involving the histone lysine N-methyltransferase KMT2A (MLL), and mutations affecting TET2, the enzyme responsible for DNA hydroxymethylation (see book ▶ Chap. 1 of Wutz and ▶ Chap. 3 of Paro). The mutual exclusion of *DNMT3A* mutations and these genomic alterations suggests a role in similar epigenetic pathways. In contrast, mutations of *DNMT3A* in AML co-occur with mutations of nucleophosmin (NPM1) and the internal tandem duplication in the receptor tyrosine kinase FLT3 gene (*FLT3*^{ITD}). Sixty percent of patients with *DNMT3A* mutations carry an *NPM1* mutation, whereas only 13% of patients with wild-type *DNMT3A* harbor an *NPM1* mutation (Yang et al. 2015). Similarly, *FLT3*^{ITD} mutations are specifically enriched in patients with *DNMT3A* mutations. When patients with *FLT3*^{ITD} mutations were classified according to *DNMT3A* status, those carrying a *DNMT3A* mutation had a significantly worse outcome, and a high relapse rate. Recent extensive genomic and epigenomic sequencing data suggested that the occurrence of all three mutations is non-random, and that *NPM1*^{mut}/*FLT3*^{ITD}/*DNMT3A*^{mut} AML is a distinct entity (Cancer Genome Atlas Research et al. 2013). However, how these mutations act together to cause leukaemia remains elusive. HSCs with *DNMT3A* mutations persist even after chemotherapy and during relapse, indicating that mutant HSCs might represent a reservoir for the re-evolution of the disease in a relapse.

8.2.4.2 Mutations of Ten-Eleven Translocation 2 (TET2)

Somatic mutations at the Ten-eleven translocation 2 (*TET2*) gene have frequently been identified in a wide variety of hematologic malignancies, including AML, myelodysplastic syndrome (MDS), chronic myelomonocytic leukemia (CMML), and myeloproliferative neoplasms (MPN). In particular, somatic deletions and inactivating mutations in *TET2* were identified in 10–20% of the MPN and MDS cases and in 7–23% of the AML cases. TET2 belongs to a family of three proteins (TET1, TET2, and TET3) that catalyze the successive oxidation of 5-methylcytosine (5meC) to 5-hydroxymethylcytosine (5hmC), 5-formylcytosine (5fC), and 5-carboxylcytosine (5caC), finally resulting in the replacement of 5meC by an unmodified cytosine (see book ► Chap. 1 of Wutz). *TET2* mutations associated with the disease are largely loss-of-function mutations that impair the enzymatic activity, resulting in the failure of the 5mC-to-5hmC conversion, and eventually impairment of DNA demethylation. Studies in mice have shown that *Tet2* loss leads to increased hematopoietic stem cell self-renewal and progressive myeloproliferation (the expansion of a multipotent hematopoietic progenitor stem cell), with features characteristic of human CMML. Thus, it appears that impaired 5meC-to-5hmC conversion confers clonal dominance to HSCs and exerts differentiation pressure toward the myeloid lineage. As in the case of *DNMT3A*, mutations in *TET2* are considered pre-leukaemic as the mutation *per se* does not induce haematologic malignancies but rather does so in conjunction with other driver mutations. The pool of pre-leukaemic HSCs containing mutations at epigenetic regulators, such as *TET2* or *DNMT3A*, has been considered a cellular reservoir that should be targeted therapeutically for more durable remissions.

Genome-wide methylation analyses revealed CpG hypermethylation of many active enhancer elements in human AML patients with *TET2* mutations and similar results were obtained using a *Tet2*-dependent leukemia mouse model. The loss of 5hmC and gain of methylation at these enhancer elements lead to decreased H3K27ac and downregulation of neighboring genes, including tumor suppressor genes.

TET enzymes are Fe(II)/α-ketoglutarate (αKG)-dependent dioxygenases. Isocitrate dehydrogenases IDH1, IDH2, IDH3 are the enzymes that generate αKG. Genome-wide sequencing studies identified mutations in *IDH1* and *IDH2* in several cancers, including AML. *IDH1/2* mutations gain the function to generate the novel oncometabolite D-2-hydroxyglutarate (D2HG) from αKG. D2HG inhibits αKG-dependent enzymes such as TET2. In AML, *TET2* and *IDH1/2* mutations are mutually exclusive, suggesting that the IDH1/2-TET2 pathway acts to suppress AML. The role and function of αKG and IDH in physiological and pathological conditions is described in book ► Chap. 9 of Santoro.

8.2.5 Epigenetic Inhibitors of DNA Methyltransferases in Cancer Therapy

Two inhibitors of DNA methyltransferases, 5-azacytidine (azacitidine) and 5-aza-2′-deoxycytidine (decitabine), have been approved for clinical use in haematological malignancies such as MDS, AML, and CMML (◘ Fig. 8.4). Azacitidine and

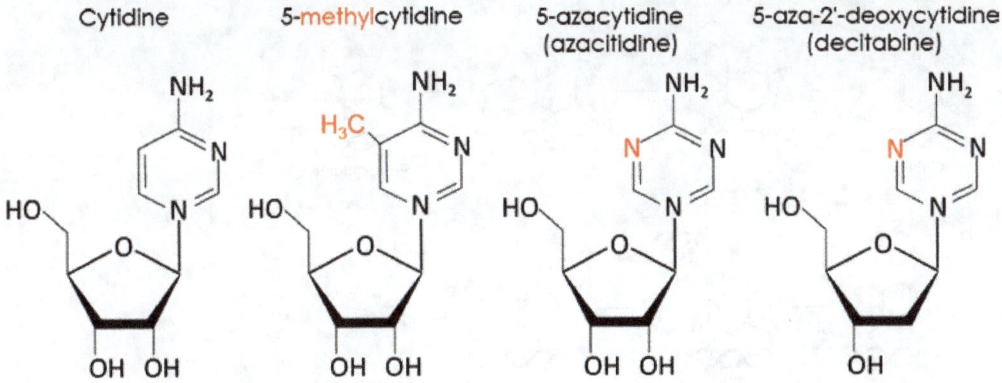

Fig. 8.4 DNA methyltransferase inhibitors. Chemical structure of cytidine nucleoside and DNA methyltransferases inhibitors azanucleosides

decitabine are nucleotide analogues. Azacitidine can be incorporated into RNA and DNA whereas decitabine only into DNA. In particular, these nucleotide analogues can be incorporated during DNA replication instead of cytosine.

Maintenance of DNA methylation occurs after the passage of the replication fork where hemi-methylated DNA is generated that is then methylated at the opposite strand by DNMT1 (see book ▶ Chap. 1 of Wutz). The reaction of DNMTs comprises the attack at the C6 position of the pyrimidine ring of cytosine, leading to the intermediate formation of a covalent DNMT-DNA complex, followed by formation of 5mC (Fig. 8.5). In the case of azacitidine or decitabine, however, the replacement of the carbon with an aza-group impairs the transfer of the methyl-group to the C-5 position (Fig. 8.5). Consequently, the reaction arrests, the DNMT remains covalently trapped on the DNA, and the restoration and maintenance of methylation after the passage of the replication fork is impaired. Thus, treatment with DNMT inhibitors leads to a passive DNA demethylation process (see book ▶ Chap. 1 of Wutz).

Azacitidine is the first therapy to have demonstrated a survival benefit for patients with MDS (Fenaux et al. 2009). However, the molecular mechanisms governing the impressive responses seen in MDS are largely unknown. It is important to note that, in addition to the demethylating effect of these compounds, also the trapping of DNMT proteins on the DNA might indirectly contribute to an anti-cancer effect. Recently, it has also been shown that treatment with DNMT inhibitors induces immune signaling in cancer cells through the upregulation of a wide range of genes implicated in immune signaling and increased numbers of immune cells in the tumor microenvironment. Transcriptional silencing mediated by DNA methylation is also related to the recruitment of co-repressors and deacetylation or methylation of histone marks at promoters and enhancers. Accordingly, the combinatorial use of DNMT and histone deacetylases (HDAC) inhibitors (described in the next paragraph) was shown to enhance reactivation of aberrantly silenced genes in tumor cells and cause reductions in tumor burden.

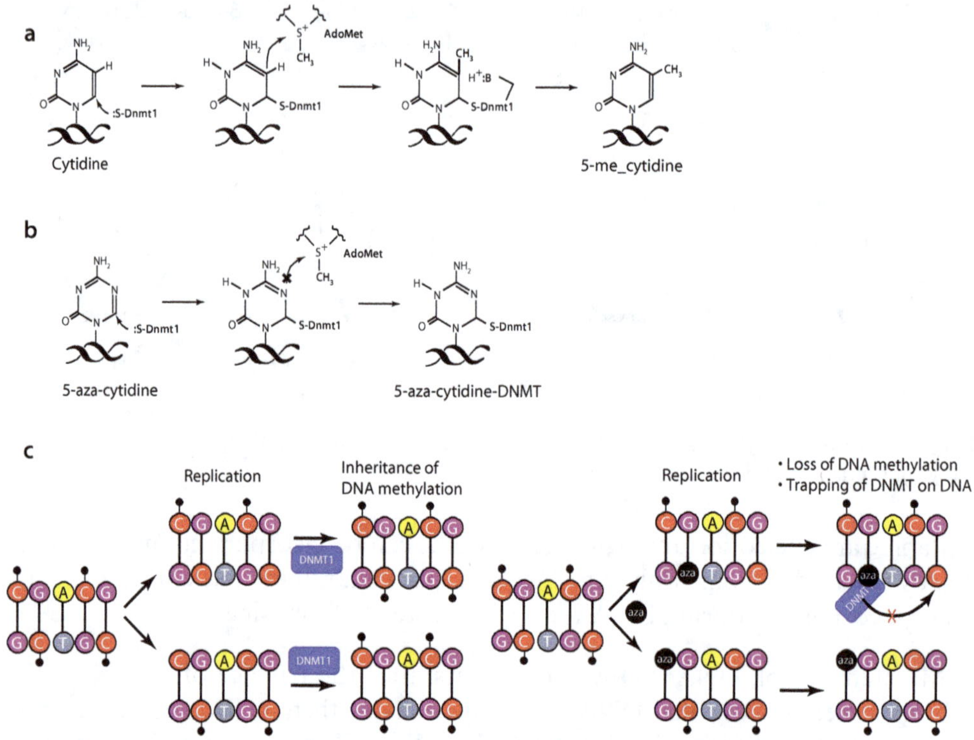

Fig. 8.5 Scheme showing the action of DNMT inhibitors. **a** The reaction of DNMTs comprises the attack at the C6 position of the pyrimidine ring of cytosine, leading to the intermediate formation of a covalent DNMT-DNA complex followed by formation of 5mC. Methylation of the C5 position induces a shift of electrons and releases the enzyme. **b** In the case of azacitidine or decitabine, the replacement of the carbon with an aza-group impairs the transfer of the methyl-group to the C-5 position. Consequently, the reaction arrests and the DNMT remains covalently trapped on the DNA. **c** Because of the semiconservative nature of DNA replication, a DNA sequence carrying symmetrical methylation marks on both strands gives rise to two hemi-methylated double strands, which can be restored to fully a methylated status by the maintenance DNA methyltransferase DNMT1. Azacitidine and decitabine can be incorporated into DNA during replication. Treatment with these compounds leads to loss of DNMT activity because the enzyme becomes irreversibly bound to the aza-residues in DNA and is no longer available for further catalysis. Loss of DNA methylation is the result of the dilution of DNA methylation during several rounds of DNA replication (passive DNA demethylation), resulting in transcriptional activation of genes previously silenced by DNA methylation

8.3 *Polycomb* Group Proteins and Cancer

In ▶ Chap. 3 "cellular memory", you have learnt that *Polycomb* group (PcG) proteins mainly function as members of two large multisubunit complexes, *Polycomb* Repressive Complex 1 and 2 (PRC1 and PRC2). PRC1 catalyzes the monoubiquitination of histone H2A at lysine 119 (H2AK119ub1) whereas PRC2 catalyzes the trimethylation of histone H3 at lysine 27 (H3K27me3) (see book ▶ Chap. 3 of Paro). Both of these post-translational modifications of histones are associated with transcriptional silencing. PcG proteins have been shown to regulate diverse biological processes during embryonic development, such as cell fate and lineage decisions, cellular memory, stem cell function, tissue homeostasis and regeneration. Because of the involvement of PcG proteins in so many key cellular processes, it should not be now surprising that alterations in the PcG machinery have frequently been found in several cancers. These changes include mutations and differential expression of the writers and erasers of H3K27me3 (8.3.1) and mutations in genes encoding histone H3 (8.3.2).

8.3.1 Alterations of PcG Activity in Cancer

Several PcG proteins are differentially expressed in tumors compared to the corresponding normal tissue (◘ Fig. 8.6). PRC2 components EZH2 and SUZ12 and PRC2 component BMI1 (see book ▸ Chap.3 of Paro) are often found overexpressed in several malignancies with an aggressive phenotype. In particular, more than half of the hormone-refractory prostate cancers, representing the late-stage cancer, exhibit amplification of the *EZH2* gene, which encodes the catalytic subunit of PRC2 responsible for H3K27me3. This amplification results in increased EZH2 protein expression that consequently alters the H3K27me3 genomic landscape. High levels of BMI1 and EZH2 predict advanced disease and a poor prognosis in several human cancer types. In particular, EZH2 level is an important factor for assessing the progression and prognosis of prostate cancers.

Cancers also exhibit several PcG missense mutations and chromosomal translocations. For example, heterozygous mutations replacing a single tyrosine in the catalytic SET domain of the EZH2 protein (Tyr641) occur in 21.7% of patients with diffuse large B-cell lymphoma and 7.2% with follicular lymphoma. This mutation alters the catalytic activity of EZH2 by inhibiting the monomethylation of H3K27 and increasing the catalytic efficiency for subsequent di- and di- to trimethylation reactions (hyperactivated EZH2 mutants). Furthermore, PcG proteins can physically interact with a number of fusion transcription factors originating from transloca-

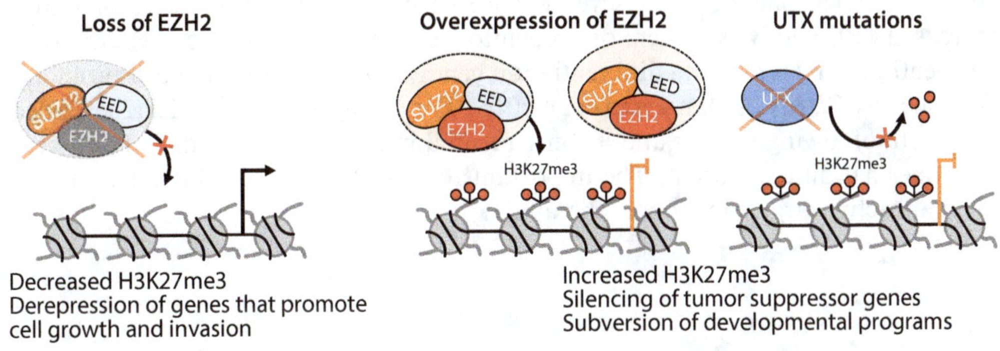

◘ **Fig. 8.6** Alterations of PcG activity in cancer. EZH2 is the catalytic subunit of PRC2 that also contains SUZ12 and EED. Loss-of-function mutations in *EZH2* gene decrease H3K27me3 levels and activate the expression of genes implicated in cancer growth and invasion, conferring a poor prognosis. Amplification of *EZH2* or mutations leading to its hyperactivation increase H3K27me3, thereby silencing tumor suppressor genes and promoting a cancer stem cell-like state. Similarly, inactivating mutations in the histone demethylase gene *UTX* cause an increase in H3K27me3

tions, such as the promyelocytic leukemia zinc finger-retinoic acid receptor α (PLZF-RARα) that, together with PRC2, induces the transcriptional silencing of target genes, thereby mediating leukemic transformation. Consistent with the critical role of H3K27me3 regulation in cancers, inactivating somatic mutations in the histone lysine demethylase gene *UTX* (KDM6A) have also been found in several cancers that consequently showed high H3K27me3 levels.

Although the role of PRC2 in cancer is mainly linked to H3K27me3 and transcriptional repression, there are cases of additional, non-classical functions. For example, in castration-resistant prostate cancers (CRPCs, i.e. cancers that do not respond to androgen deprivation therapy and hence representing the final and most aggressive stage of the disease), EZH2 acts as a coactivator for critical transcription factors. This functional switch is dependent on the phosphorylation of EZH2 at Ser21, which promotes the interaction with the Androgen Receptor (AR) and the activation of AR-target genes that are critical for disease progression (Xu et al. 2012).

Since PcG proteins are critical for maintaining stem cell-like characteristics of adult as well as embryonic stem cells (ESCs), it has been proposed that abnormally elevated levels of PcG proteins may lead to the generation and maintenance of cancer stem cells. Tumors are composed of heterogeneous populations of cells, which differ in their phenotypic and genetic features. It has been suggested that cellular heterogeneity within a tumor is organized in a hierarchical manner. In particular, evidence indicates the presence of a small subpopulation of cancer stem cells with stem cell-like characteristics that give rise to heterogeneous cancer cell lineages and undergo self-renewal to maintain their reservoir[7]. Cancer stem cells were shown to have an enhanced capacity for therapeutic resistance, immune evasion, invasion, and metastasis (Prager et al. 2019). In pluripotent ESCs, PRC1 and PRC2 bind to promoters of genes encoding regulators of developmental processes that must remain silenced to maintain ESC self-renewal. During ESC differentiation, PcG factors are removed from these genes that consequently become de-repressed, thereby setting new transcriptional programs to ensure proper differentiation. Furthermore, PcG proteins are required for cell fate determination since they restrict alternative fates once cells differentiate towards a specific cell lineage. Since PRC2 is important for balancing proliferation versus differentiation, PRC2 has been shown to promote a de-differentiated phenotype of several cancers (see also book ► Chap. 3 of Paro). Increased PRC2 activity in stem or progenitor cells might promote self-renewal over differentiation by repressing differentiation genes or genes controlling cell proliferation, such as *CDKN2A* (Laugesen et al. 2016). For example, elevated EZH2 levels led to growth of Ewing tumors and a consequential inhibition of endothelial and neuroectodermal differentiation. The important role of PcG proteins in cancer makes them attractive targets for cancer therapy.

---

7 Stem cells in normal tissues are generally defined as undifferentiated cells that have the capacity to produce the specialized end cells of a tissue as well as to self-renew to maintain their reservoir (see haematopoietic stem cells). By analogy, cancer stem cells have the ability to self-renew and to generate and re-populate a malignant cell population.

8.3.2 Mutations of Affecting Lysine 27 of Histone H3 Occur in Multiple Cancers

Consistent with the critical role of altered PRC2 activity, mutations in the substrate, lysine 27 of histone H3, have also been reported in cancers. Whole-genome sequencing of paediatric high-grade gliomas have identified gain-of-function mutations in histone H3 genes, specifically histone 3A (*H3F3A*) and histone H3B (*HIST1H3B*), encoding histone H3 variants H3.3 and H3.1, respectively (Schwartzentruber et al. 2012; Wu et al. 2012). In 78% of paediatric diffuse intrinsic pontine gliomas (DIPGs) and in 22% of non-brainstem paediatric glioblastomas (non-BS-PGs), the *H3F3A* or *HIST1H3B* genes contained a mutation leading to the substitution of lysine 27 to methionine (H3K27M). Somatic mutations at *H3F3A* that replace glycine 34 with arginine (H3G34R) were also found in 14% of non-BS-PGs.

H3K27M mutant tumors are frequently found in the thalamus, pons, and spinal cord and show poor prognosis whereas H3G34R mutations are seen in tumors of the cerebral hemispheres, suggesting a different cellular origin for these tumors. At molecular level, H3K27M inhibits the enzymatic activity of PRC2 through interaction with the EZH2 subunit and causes a profound reduction of wild-type H3K27me3 levels (Lewis et al. 2013) (■ Fig. 8.7).

The alteration in H3K27me3 levels has been thought to drive a transcriptional program that promotes tumor initiation and progression. The strong reduction of H3K27me3 in H3K27M tumors is quite remarkable since the human genome contains 16 distinct histone H3 encoding genes and only one of the two *H3F3A* genes contains the H3K27M mutation. Thus, the K27M mutation of only one histone H3 gene causes a drastic reduction of H3K27me3 at all the other wild-type histone H3s that, in diploid cells, are encoded by 31 genes. Importantly, concomitant with H3K27me3 reduction, the levels of the active H3K27ac mark increase, suggesting a switch in the epigenetic landscape of H3K27M tumors and the activation of genes linked to proliferation.

There have been several hypotheses to explain how H3K27M impairs PRC2 activity. It has been suggested that H3K27M might sequester PRC2 to chromatin and inhibit its activity. Crystal structure and *in vitro* biochemical assays showed that

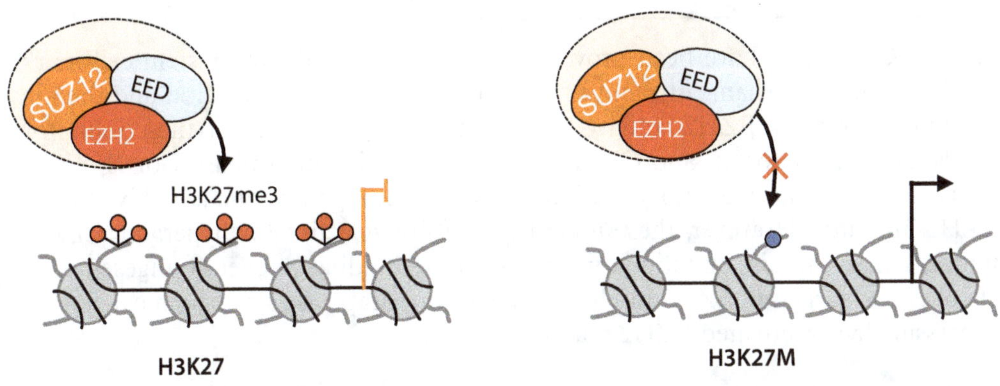

■ **Fig. 8.7** H3K27M mutations in genes encoding the histone variants H3.3 and H3.1. H3K27M mutations are recurrently found in pediatric high-grade glioma (HGG) and diffuse intrinsic pontine glioma (DIPG). K27M mutations in genes coding for histone variants H3.3 and H3.1 result in a global reduction of H3K27me3 levels, leading to derepression of PRC2 target genes

PRC2 has high binding affinity for mutant H3K27M histones, leading to the stalling of PRC2 and preventing propagation of the H3K27me3 mark. Furthermore, *in vitro* biochemical analyses revealed that PRC2 activity is inhibited by di-nucleosomes composed of H3K27M-H3K27me3. On the other hand, it has also been proposed that H3K27M might exclude PRC2 from chromatin. Indeed, H3K27M associates with chromatin regions enriched in H3K27ac whereas PRC2 is excluded from these regions. These results support a model in which H3K27M excludes PRC2 binding, which thus induces aberrant accumulation of H3K27ac at nucleosomes containing the wild-type H3 and the mutated H3K27M.

The majority of the heterotypic H3K27M-H3K27ac nucleosomes were found at actively transcribed genes and colocalized with BRD2 and BRD4, which are part of the Bromodomain (BRD) and Extra-Terminal motif (BET) protein family. BRD-containing proteins regulate gene expression primarily through recognition of histone acetyl residues, leading to the recruitment of protein complexes that modulate gene expression (see also 8.4.2). The development of inhibitors of epigenetic readers as anti-cancer agents is now an intense area of research. In particular, the selective *small*-molecule inhibitor *JQ1* binds competitively to bromodomains and impairs their binding to acetylated histones (Qi 2014). The association of BRD2 and BRD4 proteins with chromatin regions enriched in H3K27M-H3K27ac heterotypic nucleosomes suggested a potential role of BRD proteins in DIPG pathogenesis. Accordingly, treatment with JQ1 strongly inhibits the proliferation of H3K27M-DIPG cells *in vivo* and DIPG xenograft mice[8] show substantially improved survival, identifying BET proteins as potential therapeutic targets of H3K27M-DIPG. It has also been suggested that increasing H3K27me3 in H3K27M-DIPG might be an effective therapeutic strategy. This can be achieved either by enhancing PRC2 methyltransferase activity or inhibiting H3K27 demethylase activity. H3K27me3 can be demethylated by the KDM6 subfamily K27 demethylases JMJD3 and/or UTX. Accordingly, treatment with the JMJD3 K27 histone demethylase inhibitor GSKJ4 increased cellular H3K27 methylation in patient-derived H3K27M-DIPG tumor cells and showed potent antitumor activity both *in vitro* against H3K27M cells and *in vivo* against H3K27M-DIPG xenografts (◘ Fig. 8.8).

8.3.3 EZH2 Inhibitors in Cancer Therapy

Since PcG proteins have been proven to be *bona fide* and cancer stem cell markers, PcG proteins became attractive targets for both cancer prevention and therapy. Inhibitors targeting EZH2 activity have been developed. Three of these inhibitors (GSK126, Tazemetostat, and CPI-1205) are now undergoing phase I or II clinical trials. GSK126 is an S-adenosylmethionine (SAM) competitor and a highly selective EZH2 inhibitor. However, these inhibitors are not selective for hyperactive EZH2 mutants that contain mutations in the substrate-binding pocket (Laugesen et al. 2016). Thus, targeting EZH2 might be a relevant therapeutic strategy in cancers not expressing hyperactivated EZH2 mutant variants.

8 Mouse xenograft is one of the most widely used models in cancer biology. In this model, human tumor cells are transplanted, either under the skin or into the organ type in which the tumor originated, into immunocompromised mice that do not reject human cells.

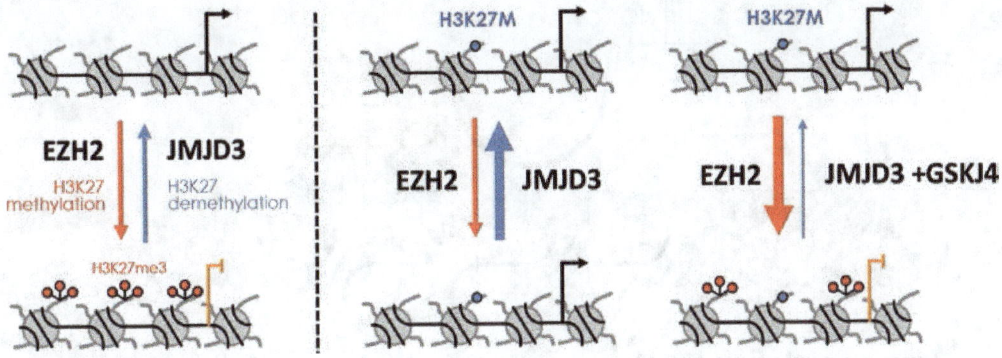

Fig. 8.8 Methylation and demethylation of H3K27. Left panel. EZH2, the catalytic component of PRC2, methylates H3K27 and promotes a transcriptionally repressed chromatin state. In contrast, JMDJD3 or UTX are demethylases that remove methyl groups from H327K, inducing an open and transcriptionally active chromatin state. Right panel. H3K27M inactivates EZH2 activity with a consequent decrease of H3K27me3 levels and transcriptional activation. Pharmacological inhibition of JMJD3 demethylase by GSKJ4 increases methylation at wild-type H3K27 and inhibits gene expression. Treatment with GSKJ4 showed potent antitumor activity in H3K27M tumor cells

8.4 Histone Acetylation and Deacetylation in Cancers

Histone acetyltransferases (HATs) are a class of enzymes that transfer acetyl groups from acetyl-CoA cofactors to lysine residues at histones (see book ► Chap. 1 by Wutz and ► Chap. 9 by Santoro). Histone acetylation is a highly reversible process since histone deacetylases (HDACs) act as erasers by removing the acetyl group. Histone acetylation is associated with active transcription, especially at promoters and enhancers, and facilitates the recruitment of co-regulators and RNA polymerase II complexes. In contrast, HDACs usually establish gene repression. As discussed in the previous section, the acetyl group can be recognized by proteins containing bromodomains (BRD) that, in turn, often recruit factors linked to transcriptional activation.

The impairment of the balance between acetylation and deacetylation can affect gene expression. This situation is often found in cancers with altered acetylation patterns. Alterations in the acetylation/deacetylation balance can originate because of the abnormal recruitment of HDACs or the reduced activity of HATs, leading to repression of tumor suppressor genes. On the other hand, it can also occur an increased HAT activity with consequent activation of oncogenes (Fig. 8.9). These mechanisms can alter the normal cell cycle, block or revert differentiation, impair apoptosis, and facilitate proliferation.

8.4.1 Alterations of Histone Acetyltransferases in Cancer

HATs are divided into three families, depending on their structural homology. Gcn5-related N-acetyltransferases (GNAT) include GCN5 and PCAF. MYST acetyltransferases include MOZ, MOF, TIP60, and HBO1. The last class consists of p300/cAMP-responsive element-binding proteins that include CBP and p300.

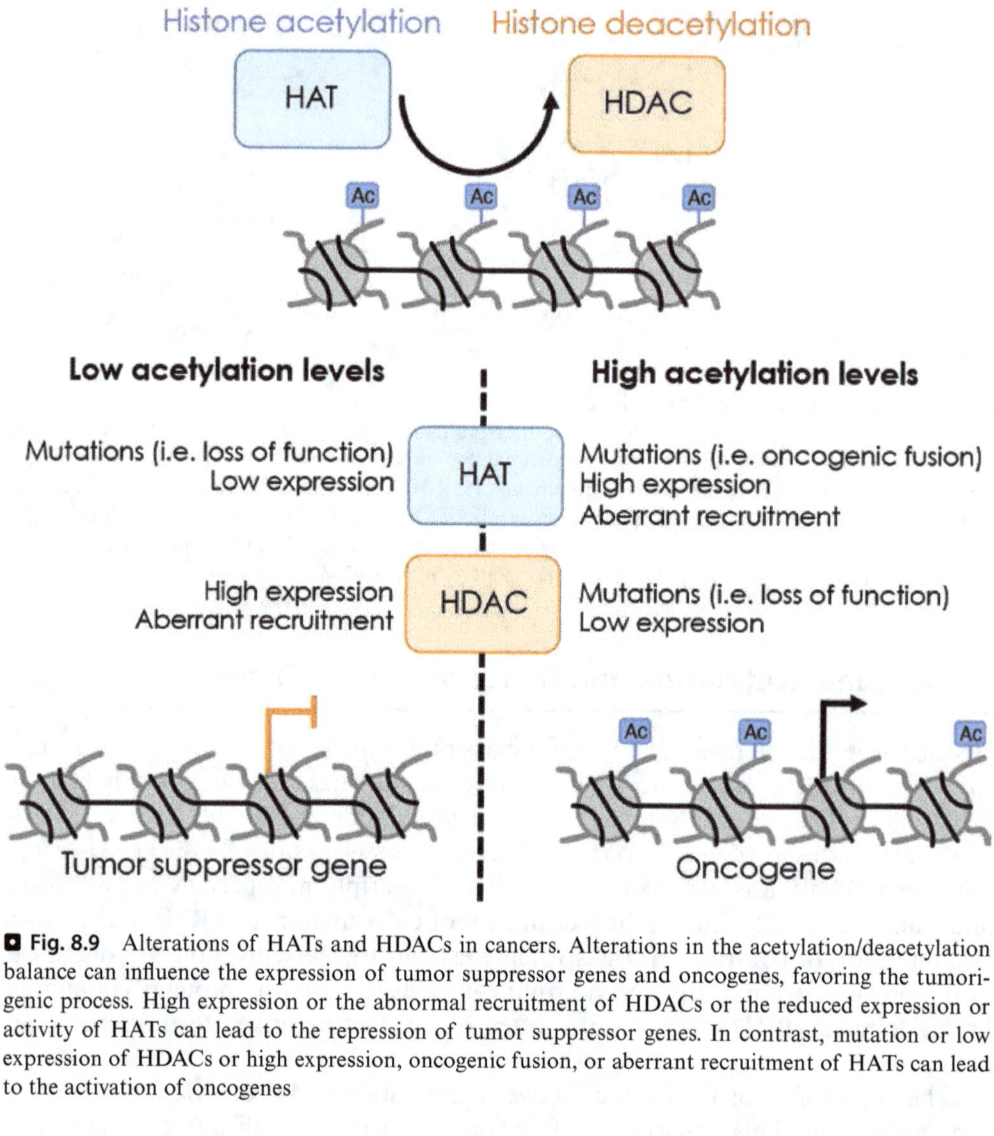

Fig. 8.9 Alterations of HATs and HDACs in cancers. Alterations in the acetylation/deacetylation balance can influence the expression of tumor suppressor genes and oncogenes, favoring the tumorigenic process. High expression or the abnormal recruitment of HDACs or the reduced expression or activity of HATs can lead to the repression of tumor suppressor genes. In contrast, mutation or low expression of HDACs or high expression, oncogenic fusion, or aberrant recruitment of HATs can lead to the activation of oncogenes

Appropriate acetylation within cells is important since upregulation or downregulation of HATs is associated with tumorigenesis and poor prognosis (Di Cerbo and Schneider 2013). For example, aberrant lysine acetylation mediated by CBP/p300 has been implicated in the genesis of multiple haematologic cancers. A high frequency of alterations in HAT genes has been reported in small-cell lung cancers, an aggressive lung tumor subtype with poor prognosis, and non-Hodgkin B-cell lymphomas. These mutations are most often point mutations in proximity to the HAT catalytic domain, resulting in loss of enzymatic activity. In acute lymphoid leukemia (ALL), a significant percentage of patients (18.3%) have mutations in the HAT domain of CBP. These mutations were shown to impair histone acetylation and transcriptional regulation of CBP targets. Since several of these mutations acquired at relapse were detected in subclones at diagnosis, it was suggested that they may confer resistance to therapy.

Genes encoding HATs can also act as oncogenes. Although less frequent, it has been reported that chromosomal translocations involving HAT genes can generate chimeric proteins that retain HAT catalytic activity and bromodomains. Beside these cases, HATs can also be involved in the regulation of oncogenic fusions. For example, AML1-ETO, the most frequent fusion protein in AMLs, is acetylated by p300. This acetylation is essential to promote self-renewal and induce leukemogenesis. These examples indicate that because of the various cellular functions linked to HATs and their specific substrates (histones but also non-histone proteins), they can act as either tumor suppressors or oncogenes, depending on the cellular or molecular context and cancer type.

8.4.2 Acetyl-Lysine Recognition Proteins and Cancer

The bromodomain (BRD) is a conserved protein module (reader) that recognizes acetyl-lysine. Depending on the structure, each BRD has preference for different acetylated histones. BRD-containing proteins are often implicated in the regulation of transcription by the recruitment of different molecular partners. HATs, such as PCAF, GCN5, p300, and CBP also contain a BRD.

As discussed earlier, an important class of BRD-containing proteins is the BET protein family, including BRD2, BRD3, BRD4, and BRDt. BET proteins are critical mediators of transcriptional activity through the recognition of acetylated histones and the recruitment of cofactors for gene activation (◘ Fig. 8.10).

BRD4 and BRD2 play an important role in transcription elongation of genes controlling cell proliferation by recruiting the positive transcription elongation factor complex (P-TEFb) to acetylated chromatin through their BRDs. BET proteins were shown to play important roles in tumorigenesis, for instance BRD4, which is required for the maintenance of AML with sustained expression of Myc. In breast cancer, BRD3/4 interacts with the histone H3K36 methyltransferase WHSC1 and promotes the expression of estrogen receptor alpha (ERα), thereby contributing to

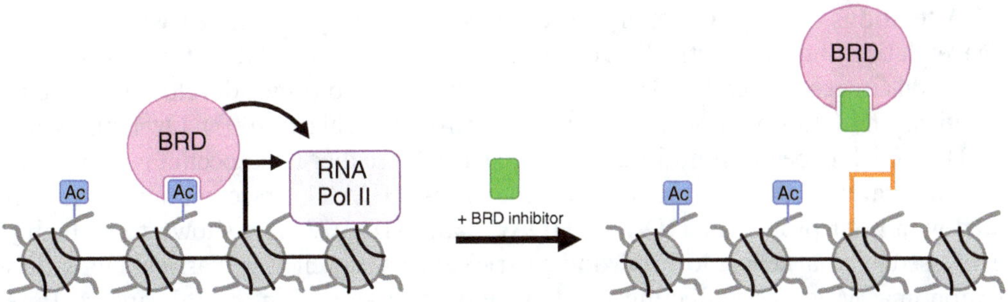

◘ **Fig. 8.10** Targeting readers of histone acetylation to modify gene expression. The bromodomain (BRD) is a conserved protein module that recognizes acetyl-lysine (reader). BRD-containing proteins are often implicated in the activation of transcription by the recruitment of different molecular partners. BRD inhibitors prevent the interaction between the BRD and the acetyl group, causing the downregulation of genes including some that may play critical roles in cancer

tamoxifen[9] resistance in ER-positive breast cancers. The fact that BRDs are a potentially druggable target has encouraged the discovery and development of several small-molecule inhibitors in recent years (◘ Fig. 8.10).

JQ1 and I-BET762 are two representative inhibitors of BET family proteins that competitively bind to BRD4 (Filippakopoulos et al. 2010). The efficacy of JQ1 was initially tested in NUT midline carcinoma (NMC) cells that express BRD4 fused to the nuclear protein in testis (NUT). The BRD4-NUT oncoprotein contributes to carcinogenesis by interfering with epithelial differentiation. Treatment of NMC cell lines and xenograft models with JQ1 was shown to displace BRD4-NUT from chromatin, inducing differentiation and specific anti-proliferative effects. To date, the efficacy of JQ1 was demonstrated in many other malignancies, including hematological malignancies and a variety of solid tumors, such as glioblastoma, medulloblastoma, hepatocellular carcinoma, colon cancer, pancreatic cancer, prostate cancer, lung cancer, and breast cancer (Perez-Salvia and Esteller 2017). Currently, there is a lot of effort in generating inhibitors targeting BRDs of non-BET proteins, including the BRD of CBP and p300. To date, many inhibitors targeting BRD proteins have been investigated in clinical trials.

8.4.3 Alterations of Histone Deacetylases in Cancer

In humans, the genome encodes 18 histone deacetylases (HDACs). HDACs can be divided into four classes, based on their primary homology to yeast HDACs. Class I HDACs include HDAC1, HDAC2, HDAC3, and HDAC8. Class II HDACs are represented by HDAC4, HDAC5, HDAC6, HDAC7, HDAC9, and HDAC10. Class III HDACs belong to the Sirtuin 2 (Sir2) family, which is composed of SIRT1, SIRT2, SIRT3, SIRT4, SIRT5, SIRT6, and SIRT7. Class IV contains only HDAC11. Classes I, II, and IV are Zn^{2+}-dependent HDACs whereas Sir2-like proteins (sirtuins) are nicotinamide adenine dinucleotide (NAD^+)-dependent HDACs (see also book ► Chap. 9 by Santoro). HDACs can also deacetylate non-histone proteins. Some HDACs are subunits of complexes, such as Sin3 and NuRD. The association with these complexes was shown to increase they catalytic activity.

Alterations of HDAC activity in cancers is generally associated with aberrant deacetylation and the inactivation of tumor suppressor genes. Altered expression and mutations in genes encoding HDACs have been linked to tumor development due to inactivation of tumor suppressor genes (Ropero and Esteller 2007). Overexpression of HDAC1 has been found in gastric, breast, pancreatic, hepatocellular, lung, and prostate carcinomas and, in most of the cases, HDAC1 up-regulation is associated with poor prognosis. HDAC1, HDAC2, and HDAC3 were shown to be highly expressed in renal cell, colorectal, and gastric cancers as well as in classical Hodgkin's lymphoma. Mutations affecting *HDACs* were also reported in several cancers. Loss

9 Tamoxifen is a hormone therapy drug used to treat breast cancer that are estrogen sensitive or estrogen receptor positive. Tamoxifen blocks the binding of estrogen receptors to estrogen, thereby the cells cannot be stimulated to divide and grow.

of HDAC2 protein expression in sporadic carcinoma was associated microsatellite instability. The best-known mechanism showing the contribution of HDACs in cancer is their interaction with oncogenic fusion proteins. In acute promyelocytic leukemia (APL), fusion proteins containing RAR-PML and RAR-PLZF bind to retinoic acid-responsive elements (RAREs) and recruit HDAC-containing complexes. This action prevents the binding of the retinoic acid receptor and represses the expression of genes implicated in the differentiation of myeloid cells. Furthermore, the fusion of ETO or TEL to AML1 converts a transcriptional activator to a constitutive transcriptional repressor through recruitment of a HDAC complex, thereby repressing the expression of genes critical to differentiation and promoting acute myeloid leukemia.

8.4.4 HAT and HDAC Inhibitors in Cancer Therapy

Aberrant HAT activity in cancers can be targeted by using HAT inhibitors. An example is provided by bi-substrate inhibitors for PCAF, p300, and TIP60 that mimic two substrates of HATs: the cofactor acetyl coenzyme A (Ac-CoA) and a peptide resembling the lysine substrate. However, these molecules are not membrane-permeable. Several synthetic compounds have been designed to selectively target HATs, such as A-485, a potent, selective, and drug-like catalytic inhibitor of p300 and CBP (Lasko et al. 2017). As described in 8.4.3, targeting of the histone acetylation readout through inhibition of the readers of histone acetylation (i.e. inhibitors targeting BRD proteins) is an attractive strategy for the pharmacological treatment of malignancies. Accordingly, BRD inhibitors have been investigated in clinical trials, whereas there currently are no clinical trials with HAT inhibitors.

HDAC family members are attractive targets for drug design and a variety of HDAC-based combination strategies have been developed for the treatment of cancers (Ropero and Esteller 2007). The inhibition of HDACs has shown promising antitumor effects. Several classes of natural and synthetic HDAC inhibitors have been identified, such as butyrates, hydroxamic acid, benzamides, and cyclic tetrapeptides. Trichostatin A (TSA) is a natural compound that inhibits the activity of HDACs and induces cancer cell differentiation and apoptosis. SAHA (vorinostat) is a pan-HDAC competitive inhibitor that structurally belongs to the group of hydroxamic acids. SAHA was the first FDA-approved HDAC inhibitor and is clinically effective in the treatment of refractory, primary cutaneous T-cell lymphomas. Specific HDAC inhibitors have also successfully been generated. SHI-1:2 is a benzamide inhibitor that shows HDAC1/HDAC2-specific inhibitory activity. Several HDAC inhibitors are currently in clinical trials and some of them have already been approved for disease treatment. As described above, mutations affecting HDACs were reported in several cancers. Mutations causing loss of HDAC2 protein expression and/or enzymatic activity rendered cells more resistant to the usual antiproliferative and pro-apoptotic effects of histone deacetylase inhibitors. Thus, the mutational status of HDAC encoding genes in individual cancers should be taken into consideration for pharmacogenetic treatments.

8.5 Chromatin Remodeling Factors and Cancer

Chromatin remodeling factors are multi-subunit complexes that use the energy of ATP hydrolysis to reposition, eject, slide, or alter the composition of nucleosomes. These processes play key roles in transcription by enabling access of DNA-binding proteins and the transcriptional machinery to DNA in order to facilitate expression. In eukaryotes, four families of chromatin remodeling complexes have been characterized: the switching defective/sucrose non-fermenting (SWI/SNF) family, the imitation-switch (ISWI) family, the nucleosome remodeling and histone deacetylase complex (NuRD), and the inositol 80 (INO80) family. These complexes contain distinct types of catalytic ATPases and associate with several factors that specify targets and contribute to diverse biological processes, including the regulation of gene expression. An extensive description of the activities of these chromatin remodeling factors can be found in book ▶ Chap. 3 of Paro. Here, the alterations of these complexes in cancer and their functional consequences are described.

8.5.1 SWI/SNF Complexes and Cancer

In mammalian cells, the SWI/SNF complexes are generally grouped into two types of complexes: BAF complexes containing the ATPase BRG1 or BRM and several alternate core subunits, such as ARID1A or ARID1B, and PBAF (polybromo-associated BAF) complexes containing the BRG1 ATPase associated with factors like PBRM1 and ARID2. Cancer genome sequencing studies have revealed that >20% of all cancers harbor mutations in SWI/SNF-encoding genes (Valencia and Kadoch 2019). Evidence for a driving role of BAF complex alterations in cancer first came from the identification of mutations at the *SMARCB1* locus encoding a subunit shared by both SWI/SNF BAF and PBAF complexes. The *SMARCB1* gene undergoes biallelic inactivation in ~98% of malignant rhabdoid tumors (MRTs), a highly aggressive paediatric cancer. Loss of *ARID1A* represents a commonly mutated gene in ovarian, endometrial, gastric, pancreas, breast, brain, prostate, lung, and liver cancers. The large majority of ARID1A mutations are frameshift indels or nonsense point mutations resulting in loss of the protein.

SWI/SNF complexes were suggested to oppose epigenetic silencing by PcG proteins. For example, in *Drosophila*, PcG proteins were shown to maintain repression of *Hox* genes during embryogenesis, while the SWI/SNF complex promotes *Hox* gene activation. Studies in MRT showed that loss of SMARCB1 destabilizes SWI/SNF complexes on chromatin, so they are unable to oppose PcG-mediated repression at bivalent promoters required for differentiation (Nakayama et al. 2017). Intriguingly, while core components of the SWI/SNF complex are frequently inactivated, PcG proteins are frequently overexpressed in cancers. For example, *EZH2* is often overexpressed in ovarian clear cell carcinomas (OCCCs) whereas *ARID1A* is mutated in ~57% of OCCCs. The link between PcG and SWI/SNF complexes is further supported by studies showing that the inhibition of EZH2 acts in a synthetic lethal manner in *ARID1A*-mutated ovarian cancer cells and the *ARID1A* mutational status correlates with the response to EZH2 inhibitors (Bitler et al. 2015).

Gain-of-function perturbations of SWI/SNF subunits have also recently been discovered. Human synovial sarcoma show a recurrent chromosomal translocation, t(X;18)(p11.2;q11.2), that fuses the *SS18* gene on chromosome 18 to one of three closely related genes on the X chromosome, *SSX1*, *SSX2*, and, rarely, *SSX4*, resulting in in-frame fusion proteins. SS18 is a subunit of BAF complexes. When *SS18* is fused to *SSX*, a protein is produced that alters the SWI/SNF complex by evicting the SWI/SNF subunit SMARCB1. This altered SWI/SNF complex is redirected to other genomic loci, relieving H3K27me3 repression at genes critical to cancer. One of these genes encodes the transcription factor *Sox2* that is essential for proliferation in synovial sarcoma. Because of the important role of SWI/SNF complexes in cancer, several compounds targeting SWI/SNF subunits have been developed. For example, the bromodomain of the SWI/SNF subunit BRD7 and BRD9 has been targeted by specific chemicals and ligand base degrader compounds, which were shown to attenuate the proliferation of tumor cells with SWI/SNF mutations (Remillard et al. 2017).

8.5.2 ISWI Complexes and Cancer

Mammals have two homologs of ISWI, SNF2L (encoded by the *SMARCA1* gene) and SNF2H (encoded by the *SMARCA5* gene). In mammals, five ISWI complexes have been identified: ACF, CHRAC, WICH, RSF, and NoRC. ISWI complexes are typically heterodimers and composed of an ISWI protein (SNF2H or SNF2L) and another subunit that provides additional specificity through targeting and recruitment of additional chromatin regulators.

RSF1 is an ISWI complex composed of SNF2H and the remodeling and spacing factor 1 (RSF1). This complex can reposition the nucleosome for transcriptional regulation. RSF1 is overexpressed in multiple types of tumors, including breast cancer, and correlates with their aggressiveness in terms of tumor size and stage. Overexpression of RSF1 was associated with the amplification of the 11q13.5 chromosomal region. Elevated expression of RSF1 was shown to promote the expression of genes regulated by NF-kB, including some involved in the evasion of apoptosis and inflammation, which is necessary for the development of chemoresistance in ovarian cancer cells (Yang et al. 2014).

BAZ2A (also known as TIP5) interacts with SNF2H to form the nucleolar remodeling complex NoRC (Bersaglieri and Santoro 2019). BAZ2A was initially identified as the repressor of ribosomal rRNA genes in healthy and differentiated cells. In prostate cancer, BAZ2A is highly expressed. Alterations in BAZ2A levels are not caused by somatic structural or sequence variations but more likely by post-transcriptional misregulation involving loss of microRNA miR-133a. In prostate cancer, BAZ2A silences the expression of numerous genes that are frequently repressed in metastatic prostate cancers. Data have also shown that BAZ2A is required for the initiation of prostate cancer driven by the loss of *PTEN*, the most commonly lost tumor suppressor gene in primary prostate cancer (Pietrzak et al. 2020). BAZ2A overexpression was shown to be tightly associated with a prostate cancer subtype displaying CIMP and prostate cancer recurrence. Thus, BAZ2A might serve as a useful marker for metastatic potential in prostate cancer.

8.5.3 The NuRD Complex and Cancer

NuRD (Mi-2) is macromolecular protein complex that combines chromatin remodeling with protein deacetylase activity (Hoffmann and Spengler 2019). The remodeling subcomplex consists of an ATPase (chromodomain helicase DNA-binding protein 3/4/5; CHD3/4/5) whereas the histone deacetylase activity is attributed to HDAC1/2. NuRD also encompasses several non-enzymatic components including methyl-CpG-binding domain 2/3 (MBD2/3), retinoblastoma-binding proteins 4/7 (RBBP4/7), metastasis-associated proteins 1/2/3 (MTA1/2/3) and GATA zinc finger domain containing proteins 2A/B (GATAD2A/B) (Allen et al. 2013).

Multiple subunits belonging to NuRD complex protein families are overexpressed in a variety of human cancers. These include MTA1/2/3, HDAC1, and HDAC2. The elevated expression of these factors was associated with gene repression and implicated in processes such as DNA damage repair and the maintenance of genomic integrity. MTA1 is overexpressed in various cancers, correlating with cancer progression and poor outcome. MTA3 has been shown to directly interact with BCL-6, a master regulator of B cell differentiation that plays a crucial role in diffuse large B cell lymphoma. The MBD3 subunit was shown to directly interact with JUN, an oncoprotein important in many malignancies. Recently, CHD4, one of the catalytic subunits of NuRD, was shown to be essential in fusion-positive rhabdomyosarcoma (FP-RMS), a rare paediatric sarcoma with a low mutational burden that exhibits features of skeletal myogenesis. The most common chromosomal translocation observed in FP-RMS is *PAX3-FOXO1*, both encoding transcription factors. This fusion generates a novel transcription factor with altered transcriptional power and target genes. Surprisingly, although CHD4 is classically defined as a repressor, in FP-RMS it acts as co-activator by interacting with super-enhancers where it generates a chromatin architecture permissive for binding of PAX3-FOXO1 and activating its downstream oncogenic program (Marques et al. 2020). These examples further indicate that because of the various cellular functions mediated by chromatin remodeling activities, their readout depend on the cellular or molecular context and cancer type.

8.5.4 The INO80 Complex and Cancer

The INO80 family of remodeling enzymes is currently composed of two classes of enzymes, Ino80 and Swr1 (Watanabe and Peterson 2010). In mammals, the chromatin-remodeling complexes of the INO80 subfamily are INO80, Snf2-related CBP activator protein (SRCAP), and p400. The mammalian INO80 complex is composed of at least 13 subunits. INO80 has been implicated in many crucial cellular functions, including transcriptional regulation, DNA replication and repair, telomere maintenance, and chromosome segregation. Subunits of the INO80 complex are frequently amplified in cancers, including lung squamous carcinoma, 50% of pancreatic cancers, and 45% of bladder cancers. Increased expression of *INO80* has been functionally associated with tumor progression. *INO80* is overexpressed in *BRAF*- and *NRAS*-mutated melanoma cancer cells. Mechanistically, INO80 interacts with super-enhancers through transcription factors such as MITF and Sox9. INO80 binding reduces nucleosome occupancy and facilitates Mediator recruitment (see also book ▶ Chap. 3 of Paro), thus promoting oncogenic transcription. Expression of *INO80* was also found upregulated in anaplastic thyroid carcinoma, an aggressive and lethal cancer with extrathyroidal invasion, distant metastasis, and resistance to conventional therapies. Interestingly, downregulation of *INO80* was shown to decrease the expression of stem cell marker genes as well as to attenuate stem cell-specific properties including the ability to form tumors. These results suggest that the role of INO80 in cancer cells is linked to its stem cell-promoting function. Accordingly, INO80 has been found to selectively activate pluripotency genes in ESCs (Wang et al. 2014), supporting the notion that genes and pathways important for ESC maintenance are often reactivated in cancer.

Take-Home Message

- Alterations of epigenetic processes affect gene expression and promote cancer initiation and progression
- Epigenetic regulators are frequently mutated in cancer
- The reversible nature of epigenetic aberrations has led to the emergence of the promising field of epigenetic therapy
- Alterations of DNA methylation are frequently observed in cancers. Tumor suppressor genes can be silenced through promoter hypermethylation. Global hypomethylation has been associated with genomic instability. Alterations of DNA methylation at imprinting control regions, with a consequent loss of imprinting, have also been linked to cancer.
- Component of the DNA methylation/demethylation machinery, DNMT3A and TET2, are frequently mutated in haematological malignancies. Alterations in the DNA methylation profile have been linked to enhanced self-renewal capacity of HSCs.
- Azacitidine and decitabine are nucleotide analogues that are used as inhibitors of DNA methyltransferases. These nucleotide analogues can be incorporated during DNA replication instead of cytosine and covalently bind DNMT1, impairing the inheritance of DNA methylation during DNA replication (passive DNA demethylation).
- Component of the PcG machinery are frequently mutated in cancers. These changes include mutations (loss of function or hyperactivation) and differential expression of the writers and erasers of H3K27me3 and mutations in genes encoding histone H3 (H3K27M). Increased PRC2 activity has been shown to promote a de-differentiated phenotype of several cancers by promote self-renewal over differentiation through the repression of differentiation genes or genes controlling cell proliferation.
- The mutation one histone H3 gene (H3K27M), identified in paediatric high-grade gliomas, exerts a *dominant* negative effect on H3K27me3.
- Expression of HATs and HDACs are frequently altered in cancers, leading to an acetylation/deacetylation imbalance that can influence the expression of tumor suppressor genes and oncogenes, favoring the tumorigenic process.
- BRD is a conserved protein module (reader) that recognizes acetyl-lysine. BRD-containing proteins are often implicated in the regulation of transcription by the

recruitment of different molecular partners. JQ1 is a small molecule inhibitors that competitively binds to the BRD of BRD4 transcription factor and displaces it by its target genes, causing differentiation and specific anti-proliferative effects. Their efficacy was demonstrated in many malignancies.

- Chromatin remodeling factors are also frequently mutated in cancers. In particular, *ARID1A*, the gene encoding a subunit of SWI/SNF BAF complex, is mutated in 57% in ovarian clear cell carcinomas. These mutations typically cause the loss of *ARID1A* protein expression and correlate with the response to EZH2 inhibitors (synthetic lethality).

References

Allen HF, Wade PA, Kutateladze TG (2013) The NuRD architecture. Cell Mol Life Sci 70(19):3513–3524. https://doi.org/10.1007/s00018-012-1256-2

Bell AC, Felsenfeld G (2000) Methylation of a CTCF-dependent boundary controls imprinted expression of the Igf2 gene. Nature 405(6785):482–485. https://doi.org/10.1038/35013100

Bersaglieri C, Santoro R (2019) Genome organization in and around the nucleolus. Cell 8(6):579. https://doi.org/10.3390/cells8060579

Bitler BG, Aird KM, Garipov A, Li H, Amatangelo M, Kossenkov AV, Schultz DC, Liu Q, Shih Ie M, Conejo-Garcia JR, Speicher DW, Zhang R (2015) Synthetic lethality by targeting EZH2 methyltransferase activity in ARID1A-mutated cancers. Nat Med 21(3):231–238. https://doi.org/10.1038/nm.3799

Brunetti L, Gundry MC, Goodell MA (2017) DNMT3A in leukemia. Cold Spring Harb Perspect Med 7(2). https://doi.org/10.1101/cshperspect.a030320

Cancer Genome Atlas Research N, Ley TJ, Miller C, Ding L, Raphael BJ, Mungall AJ, Robertson A, Hoadley K, Triche TJ Jr, Laird PW, Baty JD, Fulton LL, Fulton R, Heath SE, Kalicki-Veizer J, Kandoth C, Klco JM, Koboldt DC, Kanchi KL, Kulkarni S, Lamprecht TL, Larson DE, Lin L, Lu C, McLellan MD, McMichael JF, Payton J, Schmidt H, Spencer DH, Tomasson MH, Wallis JW, Wartman LD, Watson MA, Welch J, Wendl MC, Ally A, Balasundaram M, Birol I, Butterfield Y, Chiu R, Chu A, Chuah E, Chun HJ, Corbett R, Dhalla N, Guin R, He A, Hirst C, Hirst M, Holt RA, Jones S, Karsan A, Lee D, Li HI, Marra MA, Mayo M, Moore RA, Mungall K, Parker J, Pleasance E, Plettner P, Schein J, Stoll D, Swanson L, Tam A, Thiessen N, Varhol R, Wye N, Zhao Y, Gabriel S, Getz G, Sougnez C, Zou L, Leiserson MD, Vandin F, Wu HT, Applebaum F, Baylin SB, Akbani R, Broom BM, Chen K, Motter TC, Nguyen K, Weinstein JN, Zhang N, Ferguson ML, Adams C, Black A, Bowen J, Gastier-Foster J, Grossman T, Lichtenberg T, Wise L, Davidsen T, Demchok JA, Shaw KR, Sheth M, Sofia HJ, Yang L, Downing JR, Eley G (2013) Genomic and epigenomic landscapes of adult de novo acute myeloid leukemia. N Engl J Med 368(22):2059–2074. https://doi.org/10.1056/NEJMoa1301689

Di Cerbo V, Schneider R (2013) Cancers with wrong HATs: the impact of acetylation. Brief Funct Genomics 12(3):231–243. https://doi.org/10.1093/bfgp/els065

Feinberg AP, Vogelstein B (1983) Hypomethylation distinguishes genes of some human cancers from their normal counterparts. Nature 301(5895):89–92

Fenaux P, Mufti GJ, Hellstrom-Lindberg E, Santini V, Finelli C, Giagounidis A, Schoch R, Gattermann N, Sanz G, List A, Gore SD, Seymour JF, Bennett JM, Byrd J, Backstrom J, Zimmerman L, McKenzie D, Beach C, Silverman LR, International Vidaza High-Risk MDSSSG (2009) Efficacy of azacitidine compared with that of conventional care regimens in the treatment of higher-risk myelodysplastic syndromes: a randomised, open-label, phase III study. Lancet Oncol 10(3):223–232. https://doi.org/10.1016/S1470-2045(09)70003-8

Filippakopoulos P, Qi J, Picaud S, Shen Y, Smith WB, Fedorov O, Morse EM, Keates T, Hickman TT, Felletar I, Philpott M, Munro S, McKeown MR, Wang Y, Christie AL, West N, Cameron MJ, Schwartz B, Heightman TD, La Thangue N, French CA, Wiest O, Kung AL, Knapp S, Bradner JE

(2010) Selective inhibition of BET bromodomains. Nature 468(7327):1067–1073. https://doi.org/10.1038/nature09504

Fraga MF, Herranz M, Espada J, Ballestar E, Paz MF, Ropero S, Erkek E, Bozdogan O, Peinado H, Niveleau A, Mao JH, Balmain A, Cano A, Esteller M (2004) A mouse skin multistage carcinogenesis model reflects the aberrant DNA methylation patterns of human tumors. Cancer Res 64(16):5527–5534. https://doi.org/10.1158/0008-5472.CAN-03-4061

Hoffmann A, Spengler D (2019) Chromatin remodeling complex NuRD in neurodevelopment and neurodevelopmental disorders. Front Genet 10:682. https://doi.org/10.3389/fgene.2019.00682

Lasko LM, Jakob CG, Edalji RP, Qiu W, Montgomery D, Digiammarino EL, Hansen TM, Risi RM, Frey R, Manaves V, Shaw B, Algire M, Hessler P, Lam LT, Uziel T, Faivre E, Ferguson D, Buchanan FG, Martin RL, Torrent M, Chiang GG, Karukurichi K, Langston JW, Weinert BT, Choudhary C, de Vries P, Van Drie JH, McElligott D, Kesicki E, Marmorstein R, Sun C, Cole PA, Rosenberg SH, Michaelides MR, Lai A, Bromberg KD (2017) Discovery of a selective catalytic p300/CBP inhibitor that targets lineage-specific tumours. Nature 550(7674):128–132. https://doi.org/10.1038/nature24028

Laugesen A, Hojfeldt JW, Helin K (2016) Role of the Polycomb repressive complex 2 (PRC2) in transcriptional regulation and Cancer. Cold Spring Harb Perspect Med 6(9). https://doi.org/10.1101/cshperspect.a026575

Lewis PW, Muller MM, Koletsky MS, Cordero F, Lin S, Banaszynski LA, Garcia BA, Muir TW, Becher OJ, Allis CD (2013) Inhibition of PRC2 activity by a gain-of-function H3 mutation found in pediatric glioblastoma. Science 340(6134):857–861. https://doi.org/10.1126/science.1232245

Marques JG, Gryder BE, Pavlovic B, Chung Y, Ngo QA, Frommelt F, Gstaiger M, Song Y, Benischke K, Laubscher D, Wachtel M, Khan J, Schafer BW (2020) NuRD subunit CHD4 regulates super-enhancer accessibility in rhabdomyosarcoma and represents a general tumor dependency. Elife 9. https://doi.org/10.7554/eLife.54993

Nakayama RT, Pulice JL, Valencia AM, McBride MJ, McKenzie ZM, Gillespie MA, Ku WL, Teng M, Cui K, Williams RT, Cassel SH, Qing H, Widmer CJ, Demetri GD, Irizarry RA, Zhao K, Ranish JA, Kadoch C (2017) SMARCB1 is required for widespread BAF complex-mediated activation of enhancers and bivalent promoters. Nat Genet 49(11):1613–1623. https://doi.org/10.1038/ng.3958

Perez-Salvia M, Esteller M (2017) Bromodomain inhibitors and cancer therapy: from structures to applications. Epigenetics 12(5):323–339. https://doi.org/10.1080/15592294.2016.1265710

Pietrzak K, Kuzyakiv R, Simon R, Bolis M, Bar D, Aprigliano R, Theurillat JP, Sauter G, Santoro R (2020) TIP5 primes prostate luminal cells for the oncogenic transformation mediated by PTEN-loss. Proc Natl Acad Sci U S A 117(7):3637–3647. https://doi.org/10.1073/pnas.1911673117

Prager BC, Xie Q, Bao S, Rich JN (2019) Cancer stem cells: the architects of the tumor ecosystem. Cell Stem Cell 24(1):41–53. https://doi.org/10.1016/j.stem.2018.12.009

Qi J (2014) Bromodomain and extraterminal domain inhibitors (BETi) for cancer therapy: chemical modulation of chromatin structure. Cold Spring Harb Perspect Biol 6(12):a018663. https://doi.org/10.1101/cshperspect.a018663

Remillard D, Buckley DL, Paulk J, Brien GL, Sonnett M, Seo HS, Dastjerdi S, Wuhr M, Dhe-Paganon S, Armstrong SA, Bradner JE (2017) Degradation of the BAF complex factor BRD9 by heterobifunctional ligands. Angew Chem 56(21):5738–5743. https://doi.org/10.1002/anie.201611281

Ropero S, Esteller M (2007) The role of histone deacetylases (HDACs) in human cancer. Mol Oncol 1(1):19–25. https://doi.org/10.1016/j.molonc.2007.01.001

Schwartzentruber J, Korshunov A, Liu XY, Jones DT, Pfaff E, Jacob K, Sturm D, Fontebasso AM, Quang DA, Tonjes M, Hovestadt V, Albrecht S, Kool M, Nantel A, Konermann C, Lindroth A, Jager N, Rausch T, Ryzhova M, Korbel JO, Hielscher T, Hauser P, Garami M, Klekner A, Bognar L, Ebinger M, Schuhmann MU, Scheurlen W, Pekrun A, Fruhwald MC, Roggendorf W, Kramm C, Durken M, Atkinson J, Lepage P, Montpetit A, Zakrzewska M, Zakrzewski K, Liberski PP, Dong Z, Siegel P, Kulozik AE, Zapatka M, Guha A, Malkin D, Felsberg J, Reifenberger G, von Deimling A, Ichimura K, Collins VP, Witt H, Milde T, Witt O, Zhang C, Castelo-Branco P, Lichter P, Faury D, Tabori U, Plass C, Majewski J, Pfister SM, Jabado N (2012) Driver mutations in histone H3.3 and chromatin remodelling genes in paediatric glioblastoma. Nature 482(7384):226–231. https://doi.org/10.1038/nature10833

Soejima H, Higashimoto K (2013) Epigenetic and genetic alterations of the imprinting disorder Beckwith-Wiedemann syndrome and related disorders. J Hum Genet 58(7):402–409. https://doi.org/10.1038/jhg.2013.51

Valencia AM, Kadoch C (2019) Chromatin regulatory mechanisms and therapeutic opportunities in cancer. Nat Cell Biol 21(2):152–161. https://doi.org/10.1038/s41556-018-0258-1

Wang L, Du Y, Ward JM, Shimbo T, Lackford B, Zheng X, Miao YL, Zhou B, Han L, Fargo DC, Jothi R, Williams CJ, Wade PA, Hu G (2014) INO80 facilitates pluripotency gene activation in embryonic stem cell self-renewal, reprogramming, and blastocyst development. Cell Stem Cell 14(5):575–591. https://doi.org/10.1016/j.stem.2014.02.013

Watanabe S, Peterson CL (2010) The INO80 family of chromatin-remodeling enzymes: regulators of histone variant dynamics. Cold Spring Harb Symp Quant Biol 75:35–42. https://doi.org/10.1101/sqb.2010.75.063

Wu G, Broniscer A, McEachron TA, Lu C, Paugh BS, Becksfort J, Qu C, Ding L, Huether R, Parker M, Zhang J, Gajjar A, Dyer MA, Mullighan CG, Gilbertson RJ, Mardis ER, Wilson RK, Downing JR, Ellison DW, Baker SJ (2012) Somatic histone H3 alterations in pediatric diffuse intrinsic pontine gliomas and non-brainstem glioblastomas. Nat Genet 44(3):251–253. https://doi.org/10.1038/ng.1102

Xu K, Wu ZJ, Groner AC, He HH, Cai C, Lis RT, Wu X, Stack EC, Loda M, Liu T, Xu H, Cato L, Thornton JE, Gregory RI, Morrissey C, Vessella RL, Montironi R, Magi-Galluzzi C, Kantoff PW, Balk SP, Liu XS, Brown M (2012) EZH2 oncogenic activity in castration-resistant prostate cancer cells is Polycomb-independent. Science 338(6113):1465–1469. https://doi.org/10.1126/science.1227604. 338/6113/1465 [pii]

Yang YI, Ahn JH, Lee KT, Shih Ie M, Choi JH (2014) RSF1 is a positive regulator of NF-kappaB-induced gene expression required for ovarian cancer chemoresistance. Cancer Res 74(8):2258–2269. https://doi.org/10.1158/0008-5472.CAN-13-2459

Yang L, Rau R, Goodell MA (2015) DNMT3A in haematological malignancies. Nat Rev Cancer 15(3):152–165. https://doi.org/10.1038/nrc3895

9

Biosynthetic Pathways of Cofactors

Contents

What You Will Learn in This Chapter

Most chromatin-modifying enzymes use metabolites as cofactors. Consequently, the cellular metabolism can influence the capacity of the cell to write or erase chromatin marks. This points to an intimate relationship between metabolic and epigenetic regulation. In this chapter, we describe the biosynthetic pathways of cofactors that are implicated in epigenetic and chromatin regulation and provide examples of how metabolic pathways can influence chromatin and epigenetic processes as well as their interplay in developmental and cancer biology.

9.1 Epigenetics and Metabolism

Cellular metabolism involves a set of complex and highly coordinated biochemical reactions that convert or use energy to maintain the living state of a cell. This process engages regulatory mechanisms that enable cells to sense nutrient availability and transmit the information through signaling networks. Metabolism impacts every cellular process. Nowadays, the study of metabolism is influencing all fields of biological research. Most chromatin-modifying enzymes use as cofactors important intermediates of cellular metabolism. Depending on dietary intake, metabolite concentrations can vary, and, in turn, can affect gene expression by modulating the activity of epigenetic pathways and associated chromatin modifying enzymes. Considering that the methylation of DNA and each post-translational modification (PTM) of proteins can be affected by many metabolic pathways, it is clear that the epigenome might act as a sensor of the whole metabolic network.

9.2 Acetyl-Coenzyme A (Acetyl-CoA)

Acetyl-coenzyme A (acetyl-CoA) is the cofactor of histone acetyltransferases (HATs) (◘ Fig. 9.1). The abundance of acetyl-CoA reflects the general energetic state of the cell. Biosynthesis, compartmentalization, and fluctuation of acetyl-CoA concentration change considerably in response to a series of physiological or pathological conditions. Thus, acetyl-CoA acts as metabolic sensor by signaling changes in metabolism that are then converted into gene expression states.

9.2.1 Biosynthesis of Acetyl-CoA

Acetyl-CoA comprises an acetyl moiety linked to coenzyme A, a derivative of vitamin B5 and cysteine (◘ Fig. 9.1). Acetyl-CoA is a central metabolite that interconnects multiple metabolic pathways. It can be found in two separate pools in the cell: the mitochondrial pool and the nuclear/cytosolic pool (◘ Fig. 9.2). This distinction is due to the fact that the inner mitochondrial membrane is impermeable to the highly charged acetyl-CoA molecule whereas nuclear pores allow acetyl-CoA to freely distribute between the cytosol and the nucleus.

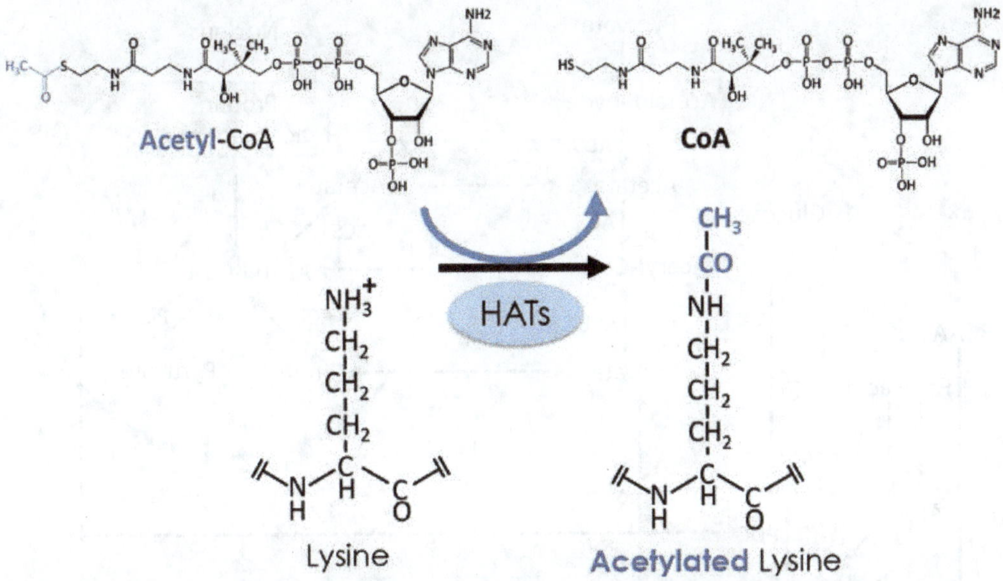

Fig. 9.1 Acetyl-coenzyme A is the cofactor of histone acetyltransferases. Acetyl-coenzyme A (acetyl-CoA) comprises an acetyl moiety linked to coenzyme A, a derivative of vitamin B5 and cysteine. The schematic shows the acetylation of the ε-amino group of lysine mediated by histone acetyltransferases (HATs), whereby the acetyl group of acetyl-CoA is transferred to the lysine

The majority of acetyl-CoA is produced and consumed in mitochondria where it allows glycolytic[1] pyruvate to enter the citric acid cycles, also known or tricarboxylic acid (TCA) cycle. In the cytosolic compartment, acetyl-CoA supports several anabolic reactions including synthesis of fatty acids and steroid and protein acetylation. In mitochondria, acetyl-CoA is produced through carboxylation of pyruvate to form acetyl-CoA, CO_2, and NADH, a reaction catalyzed by the mitochondrial pyruvate dehydrogenase complex (PDC) (Fig. 9.2). At high glucose levels, acetyl-CoA is prevalently produced during the oxidation of glucose whereas at low glucose levels, it is generated as the end-product of the ß-oxidation of fatty acids. Acetyl-CoA can also be produced by branched-chain amino acids, i.e., valine, leucine, and isoleucine, through cytosolic transamination to branched-chain α-ketoacids followed by import into mitochondria. There, a reaction catalyzed by the mitochondrial branched-chain α-ketoacid dehydrogenase generates NADH, acetyl-CoA, and other acyl-CoA thioesters. In addition to these nearly ubiquitous metabolic networks, mitochondrial acetyl-CoA can be produced by organ-specific pathways.

The nuclear-cytoplasmic level of acetyl-CoA in mammalian cells is established primarily by the combined action of two enzymes: ATP-citrate lyase (ACL) and acetyl-CoA synthetase (ACS) (Fig. 9.2). ACL converts citrate into acetyl-CoA and oxaloacetate. This cytoplasmic pool of acetyl-CoA can originate from the

1 Glycolysis is a cytoplasmatic pathway that breaks down glucose into three carbon compounds and generates pyruvate and energy in form of ATP and NADH.

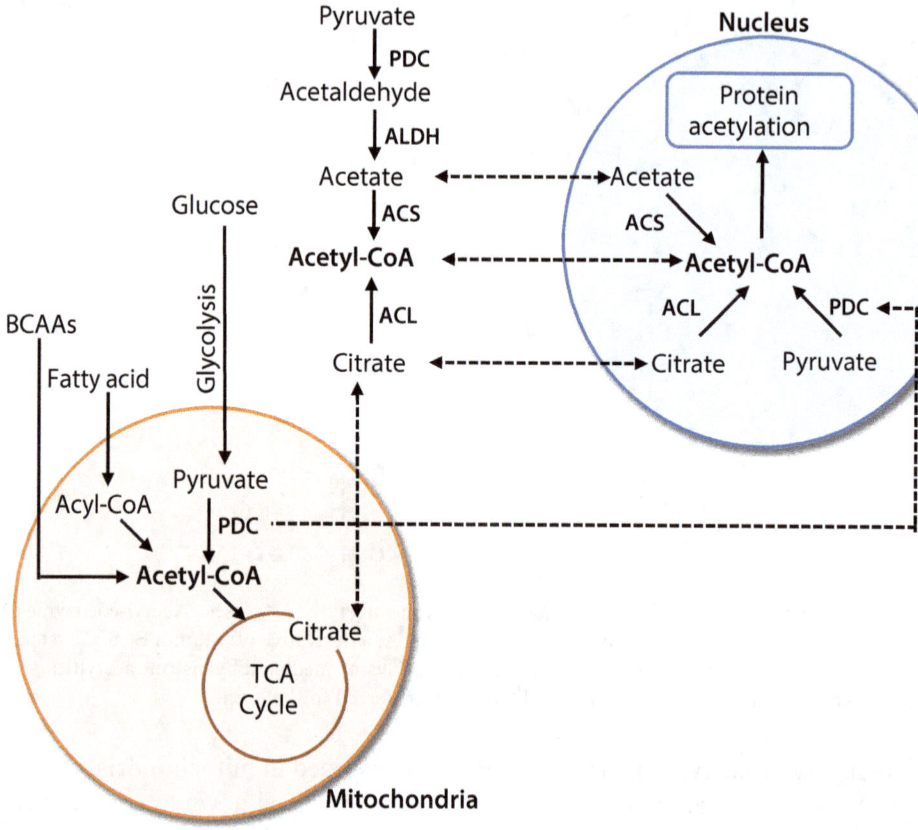

Fig. 9.2 Metabolism of acetyl-CoA in mammalian cells. Acetyl-CoA can be found in two distinct pools in the cell: the mitochondrial pool and the nuclear/cytosolic pool. The majority of cellular acetyl-CoA in mitochondria is generated and consumed in reactions involving the oxidative metabolism of glycolytic pyruvate, free fatty acids, branched-chain amino acids (BCAAs), or ketone bodies, which all converge on the citric acid cycle (TCA). In the cytosolic and nuclear compartment, acetyl-CoA is produced by ATP-citrate lyase (ACL) and acetyl-CoA synthetase (ACS). ACL converts citrate into oxaloacetate and acetyl-CoA. ACS utilizes acetate as a substrate for acetyl-CoA synthesis. These reactions support several anabolic reactions, including lipogenesis, steroidogenesis, and the synthesis of specific amino acids. PDC, ACL, and ACS are also found in the nucleus and produce acetyl-CoA therein. Their role in the nucleus mainly serves to produce acetyl-CoA for the acetylation of histones and non-histone proteins, thereby regulating gene expression

reductive carboxylation of glutamine. Cytosolic glutamine is transformed into glutamate, transported into mitochondria, and converted into α-ketoglutarate, which generates citrate within the TCA cycle. Finally, citrate can be exported back to the cytosol where is converted into oxaloacetate and acetyl-CoA by ACL. ACS utilizes acetate as a substrate for acetyl-CoA synthesis. Production of acetyl-CoA within the nucleus is also possible due to the presence, in the nucleus, of ACL and ACS. In *Saccharomyces cerevisiae*, inactivation of ACS leads to rapid histone deacetylation and transcriptional defects, underscoring a direct link between acetyl-CoA producing enzymes, chromatin regulation, and gene expression. A novel pathway for nuclear acetyl-CoA synthesis has recently been suggested by showing that PDC can translocate from mitochondria to the nucleus upon growth factor stimulation (■ Fig.9.2).

9.2.2 Acetyl-CoA as Cofactor of Histone Acetyltransferases

Acetyl-CoA is the cofactor required by histone acetyltransferases (HATs), which attach an acetyl group to the ε-amino group of specific lysine residues (◘ Fig. 9.1). Type B HATs are cytoplasmic enzymes, which modify free histones in the cytoplasm just after their synthesis[2]. Type A HATs are mainly nuclear enzymes and they are responsible for acetylation of histones and non-histone proteins in the nucleus and implicated in the epigenetic regulation of gene expression (see also book ▶ Chap. 1 of Wutz). In response to a series of physiological or pathological conditions, compartmentalization and fluctuation of acetyl-CoA concentration changes considerably and has a major impact in protein acetylation.

An important example of how acetyl-CoA synthesis responds to stimuli and affects histone acetylation and gene expression is provided by experiments studying growth of prototrophic[3] strains of yeast under a nutrient-limiting environment. Under these conditions, cells enter into synchronous and highly robust oscillations of oxygen consumption, termed yeast metabolic cycles (YMCs). During YMCs, cells go through a synchronized transition between three metabolic phases, termed oxidative (OX), reductive and building (RB), and reductive and charging (RC). Acetyl-CoA levels strongly fluctuate as a function of the YMC. In the OX phase, mitochondrial respiration peaks and is associated with the rapid induction of genes involved in growth. In the RB phase, the rate of oxygen consumption begins to decrease and cell division initiates. In the RC phase, cells enter into a stationary and quiescent phase and express many genes that are negatively correlated with increasing growth rate. In the transition from the RC to the OX phase, cells re-enter into growth. This time point is characterized by a substantial increase in intracellular acetyl-CoA levels that trigger a series of histone acetylation events catalyzed by Spt-Ada-Gcn5-acetyltransferase (SAGA, see also ▶ Chap. 2 of Paro) at genes important for growth, thereby enabling their rapid transcription and commitment to growth. The peak of acetyl-CoA at the OX/RB boundary is likely coordinated by induction of all enzymes required to convert ethanol (fermented via the consumption of glucose during the RB phase) into acetylaldehyde, into acetate, and, finally, into acetyl-CoA (Kaelin Jr. and McKnight 2013). Moreover, the link of this metabolic pathway with histone acetylation and gene expression is supported by the fact that acetyl-CoA synthase, which produces acetyl-CoA from acetate, is localized in the nucleus. In mammals, high acetyl-CoA concentrations also favor cell growth and replication at the expense of differentiation. The metabolism of embryonic stem cell (ESCs) shows the Warburg effect, a shift from oxidative phosphorylation to aerobic glycolysis that is also utilized by cancer cells[4]. Thus, ESCs produce acetyl-CoA and acetate

2 Note that the cytoplasmatic acetylation of newly synthesized histones plays a critical role in their transport and assembly into chromatin. Once incorporated into chromatin, the cytoplasmatic acetylation at these newly synthesized histone is removed by histone deacetylases.

3 Prototroph refers to an organisms or cell capable of synthesizing all its metabolites from inorganic material, without requiring organic nutrients.

4 When oxygen is limiting, cells can redirect the pyruvate generated by glycolysis away from mitochondrial oxidative phosphorylation by producing lactate (Warburg effect). This generation of lactate during anaerobic glycolysis allows glycolysis to continue, with consequent increase of Acetyl-CoA levels.

through glycolysis, however, this processing step is rapidly downregulated during differentiation. This metabolic switch causes a loss of histone acetylation during the first hours of differentiation. The link of acetyl-CoA levels and pluripotency state is supported by experiments showing that an increase of acetate, a precursor of acetyl-CoA, delays ESC differentiation and blocks early histone deacetylation whereas glycolysis inhibition leads to deacetylation and differentiation of ESCs (Moussaieff et al. 2015). Acetyl-CoA fluctuations during ESC differentiation have also been correlated with changes in the expression of threonine dehydrogenase (TDH), which converts threonine into glycine and acetyl-CoA in mitochondria, with glycine facilitating one-carbon metabolism and the synthesis of S-adenosylmethionine (SAM) for DNA and histone methylation (see ► Sect. 9.4) and acetyl-CoA feeding the TCA cycle. Another example of the link between metabolism and epigenetic gene regulation are adipocytes, cells that are the primary constituent of adipose tissue and specialized in storing energy as fat. In adipocytes, ACL, the enzyme that converts glucose-derived citrate into acetyl-CoA, is required to increase histone acetylation in response to growth factor stimulation and during differentiation, indicating that glucose availability can affect histone acetylation in an ACL-dependent manner (Wellen et al. 2009).

The local abundance of acetyl-CoA within the nucleus might also affect HAT catalytic activity and regulate gene expression by controlling the levels of histone acetylation at specific genes. HATs display a relatively high dissociation constant (KD) for acetyl-CoA and consequently local changes in acetyl-CoA concentration might affect HAT activity. Studies in yeast have shown that acetyl-CoA levels (3–30 μM) may fluctuate within a range to limit the activity of the HAT GCN5, which has a KD for acetyl-CoA of 8.5 mM. This is not the case for kinases, which use ATP as cofactor to phosphorylate, due to the elevated concentration of cellular ATP (roughly mM in cells) and their relatively high affinity for ATP (KD roughly μM). It is also worth to mention that many HATs are subject to product inhibition by free CoA, indicating that the acetyl-CoA/CoA ratio might be the relevant regulator of HAT activity instead of the absolute levels of acetyl-CoA.

The concentration of acetyl-CoA can also alter the substrate specificity of HATs. The lysine acetyltransferases CBP and p300 acetylate many common lysine residues on histones H3 and H4. However, under limiting acetyl-CoA concentrations, CBP and p300 display distinct specificity and selectivity, with p300 showing the highest specificity for H4K16, that is 10^{18}-fold higher than CBP.

9.3 Nicotinamide Adenine Dinucleotide (NAD)

Nicotinamide adenine dinucleotide (NAD) serves both as a critical cofactor for enzymes that catalyze reduction-oxidation reactions and as a co-substrate for three classes of enzymes: (i) the sirtuins (SIRTs), (ii) the adenosine diphosphate (ADP)-ribose transferases (ARTs) and poly(ADP-ribose) polymerases (PARPs), and (iii) the cyclic ADP-ribose (cADPR) synthases (◘ Fig. 9.3) (Verdin 2015).

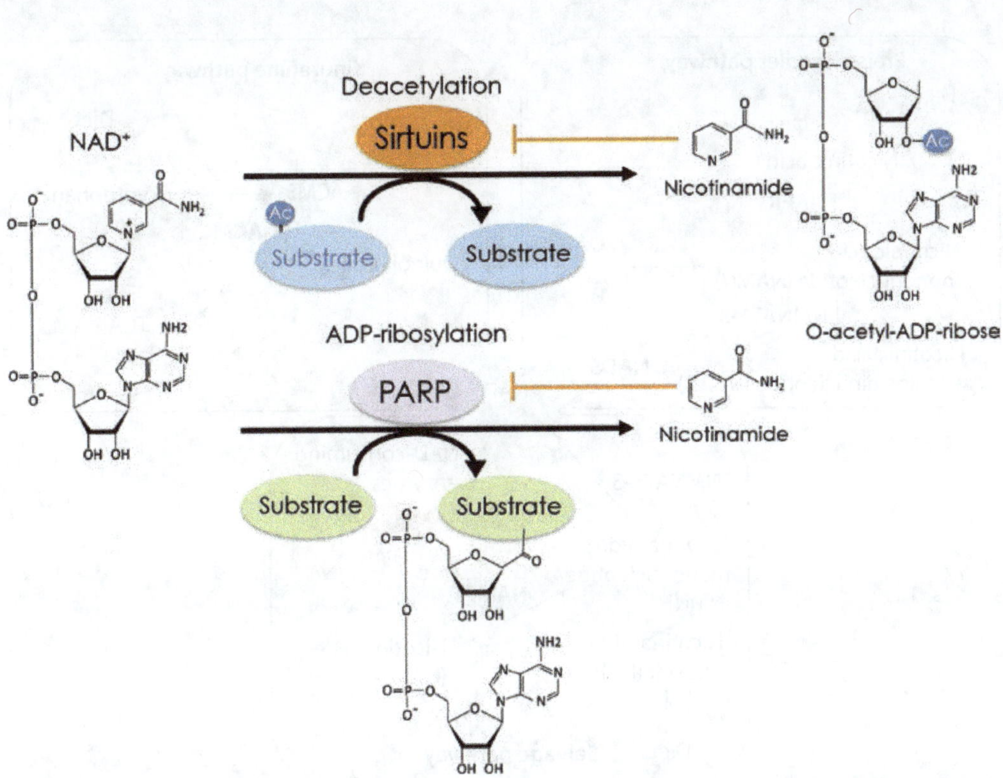

Fig. 9.3 Nicotinamide adenine dinucleotide (NAD) is a co-substrate for histone deacetylation and ADP-ribosylation. The structure of NAD in its oxidized form (NAD^+) is shown. Sirtuins remove acetyl groups from target proteins. In this reaction, NAD is cleaved into nicotinamide, and the ADP-ribose serves as the acceptor for the removed acetyl group. Adenosine diphosphate (ADP)-ribose transferases (ARTs) and poly(ADP-ribose) polymerases (PARPs) transfer the ADP-ribose moiety from NAD to a target protein, releasing nicotinamide. Nicotinamide also serves as an inhibitory factor of SIRTs, ARTs, and PARPs

9.3.1 Biosynthesis of NAD

NAD can be synthesized from diverse dietary sources, including nicotinic acid (also known as Vitamin B3, an essential human nutrient) and nicotinamide, tryptophan, and nicotinamide riboside (NR). The major dietary source of NAD is nicotinic acid. NAD levels are maintained by three independent pathways: the Preiss-Handler pathway, the kinurenine pathway, and the NAD salvage pathway (Fig. 9.4).

In the Preiss-Handler pathway, dietary nicotinic acid and the enzyme nicotinic acid phosphoribosyltransferase (NAPRT) generate nicotinic acid mononucleotide (NAMN), which is in turn converted by the nicotinamide mononucleotide adenylyltransferase (NMNAT) into nicotinic acid adenine dinucleotide (NAAD) in the presence of adenosine triphosphate (ATP). NMNAT has three forms with distinct subcellular localizations: NMNAT1 in the nucleus, NMNAT2 in the cytosol and Golgi, and NMNAT3 in the cytosol and mitochondria. The kinurenine pathway

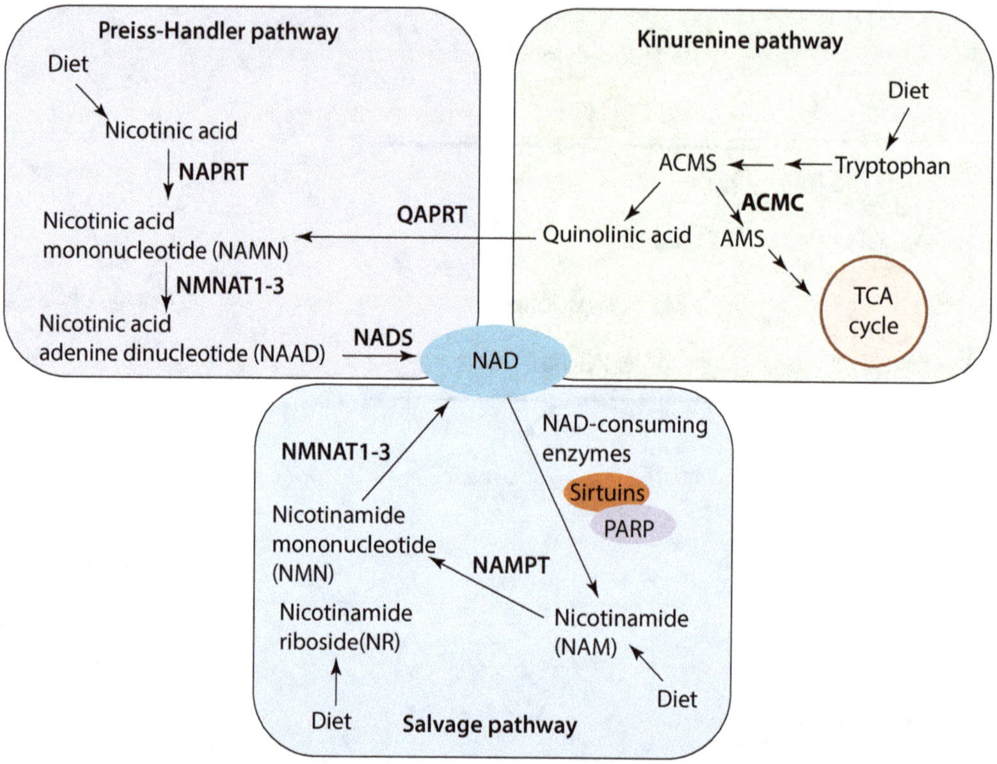

Fig. 9.4 NAD biosynthetic pathways. Three independent pathways (Preiss-Handler, kinurenine and salvage pathways) maintain NAD levels. In the Preiss-Handler pathway, NAPRT enzyme transforms nicotinic acid into NAMN, which is then converted into NAAD by NMNAT1-3. NAD is generated from NAMN by the reaction catalyzed by NAD synthase (NADS). In the kinurenine pathway, NAD is produced from tryptophan. Spontaneous condensation ACMS and rearrangement into quinolinic acid make the kinurenine pathway to converge into the Preiss-Handler pathway for the production of NAD. The NAD salvage pathway recycles the nicotinamide generated as a by-product of the enzymatic activities of NAD-consuming enzymes SIRTs, ARTs, and PARPs. NAMPT uses nicotinamide to produce NMN, which is then converted into NAD via the different NMNATs as in the Preiss-Handler pathway. Nicotinic acid phosphoribosyltransferase (NAPRT), Nicotinic acid mononucleotide (NAMN), Nicotinic acid adenine dinucleotide (NAAD), Nicotinamide mononucleotide adenylyltransferase (NMNAT1-3), 2-amino-3-carboxymuconate semialdehyde (ACMS), Nicotinamide phosphoribosyltransferase (NAMPT), Nicotinamide mononucleotide (NMN)

produces NAD from tryptophan and converges into the Preiss-Handler pathway since 2-amino-3-carboxymuconate semialdehyde (ACMS) undergoes spontaneous condensation and rearranges into quinolinic acid, which is transformed into NAMN.

The NAD salvage pathway is a key pathway for maintaining cellular NAD levels since it recycles the nicotinamide generated as a by-product of the enzymatic activities of NAD-consuming enzymes, the SIRTs, ARTs, and PARPs. Nicotinamide phosphoribosyltransferase (NAMPT) recycles nicotinamide into nicotinamide mononucleotide (NMN), which is then converted into NAD via the different NMNATs as in the Preiss-Handler pathway. As it will be discussed in ► Sect. 9.3.2, nicotinamide also serves as an inhibitory factor of SIRTs, ARTs, and PARPs (◘ Fig. 9.3). Consequently, the NAD salvage pathway leads not only to recycling of nicotinamide into NAD but also relieves nicotinamide inhibition of NAD-consuming enzymes.

Cellular NAD levels decline during the process of chronological aging and in progeroid[5] states. NAD levels decrease in animals kept under a high fat diet whereas NAD increases under exercise and caloric restriction. The relationship between NAD levels and the nutritional state of the organism has also been confirmed in studies of the circadian clock, which is encoded by a transcription-translation feedback loop that synchronizes behavior and metabolism with the light-dark cycle. Both the expression of NAMPT, the rate-limiting enzyme in the NAD salvage pathway, and the levels of NAD cycle with a 24-hour rhythm are regulated by the core clock machinery in mice.

NAMPT has been implicated in the decline of cellular NAD levels associated with aging. NAMPT expression decreases during age and forced expression of NAMPT delays senescence and substantially lengthens cell lifespan. In addition, lifelong muscle-specific *Nampt* transgene expression preserved muscle NAD levels and exercise capacity in aged mice (Frederick et al. 2016). The link between NAD levels and aging has initiated considerable efforts to manipulate NAD concentrations in therapies aimed at disease prevention and life-span extension. Pharmacological approaches able to improve NAD availability have been investigated as potential therapeutic treatments for different human disorders. NR is a NAD precursor that has the ability to cross the plasma membrane and does not inhibit the activity of sirtuins and PARPs. Diet supplementation with NR was shown to induce the mitochondrial unfolded protein response and synthesis of prohibitin proteins, rejuvenate muscle stem cells in aged mice and in a mouse model of muscular dystrophy (Zhang et al. 2016). Similarly, administration of the NAD intermediate NMN, a product of the NAMPT reaction, has been shown to enhance NAD biosynthesis and to mitigate age-associated physiological decline in mice. Supplementation of NMN was shown to suppress age-associated body weight gain, enhance energy metabolism, promote physical activity, improve insulin sensitivity and the plasma lipid profile, and ameliorate reduced eye function and other pathophysiologies (Mills et al. 2016).

9.3.2 NAD as Cofactor of Sirtuins and PARPs

NAD is a critical cofactor for SIRTs, ARTs, and PARPs. Although these three classes of enzymes use NAD, they have distinct functional roles. In the next sections, the role of SIRT and PARP enzymes will be discussed and their regulation, which is mediated by NAD levels.

9.3.2.1 Sirtuins

Sirtuins are class III histone deacylases that remove acyl groups from lysine residues on proteins in a NAD-dependent manner (◘ Fig. 9.3). NAD is cleaved between nicotinamide and ADP-ribose, and the latter serves as an acyl acceptor, generating acyl-ADP-ribose.

In mammalian cells, there are seven sirtuins, SIRT1-7, which are located in distinct cellular compartments and can, therefore, coordinate cellular responses to

5 Progeroid syndromes are a group of rare genetic disorders that showed accelerating physiological ageing, such as hair loss, short stature, skin tightness, cardiovascular diseases and osteoporosis.

caloric restriction. SIRT1, SIRT6, and SIRT7 are localized in the nucleus where they function to deacetylate histones, thereby affecting gene expression. SIRT2 is both a cytosolic and nuclear sirtuin and modulates cell cycle control. SIRT3, SIRT4, and SIRT5 are mitochondrial sirtuins and respond to caloric restriction by switching cells to favor mitochondrial oxidative metabolism (Chang and Guarente 2014).

Sirtuin function is intrinsically linked to cellular metabolism. Over the past two decades, accumulating evidence indicates that sirtuins are conserved, diet-sensitive, anti-aging proteins. Caloric restriction, a decrease in calorie intake (by 10 to 40%) without malnutrition, has been shown to increase lifespan and healthspan in all organisms in which it has been tested. Moreover, as discussed above, NAD levels decline with aging, which would lead to a reduction in sirtuin activity. Initial studies in *S. cerevisiae* showed that the increased longevity induced by caloric restriction requires the activation of the sirtuin protein silent information regulator 2 (SIR2) by NAD (Lin et al. 2000). Since then, several other studies in flies and worms have confirmed the lifespan-extending effects of SIR2 homologs, underscoring the evolutionary significance of SIR2 in extending life span. Studies in mice also revealed a role of sirtuins in the regulation of aging and longevity in mammals. Transgenic mice overexpressing SIRT1 and SIRT6 exhibit phenotypes consistent with a delay in aging.

SIRT1 functions as a chromatin regulator by deacetylating specific histone-acetylated residues (e.g., H3K9ac, H3K14ac, and H4K16ac) but also transcription factors such as TP53, NF-kB, PGC-1a, and FOXO3a. SIRT1 protein abundance is relatively stable, but its deacetylase activity depends on NAMPT to generate NAD. An important example is provided by the regulation of SIRT1 activity in circadian clock regulation. SIRT1 is recruited to the *Nampt* promoter inhibiting the expression of NAMPT, which is a key enzyme of the NAD salvage pathway, and thus contributes to the circadian synthesis of its own coenzyme. Inhibition of *Nampt* promotes oscillation of the clock gene *Per2* by releasing CLOCK:BMAL1 from suppression by SIRT1. In turn, the circadian transcription factor CLOCK binds to and up-regulates *Nampt*, thus completing a feedback loop involving NAMPT/NAD and SIRT1/CLOCK: BMAL1. Caloric restriction increases SIRT1 protein levels and induces neural activation in the dorsomedial and lateral hypothalamic nuclei. Increasing SIRT1 in the brain of transgenic mice enhances neural activity specifically in hypothalamic nuclei and promotes physical activity in response to different diet-restricting paradigms. However, moderate overexpression of SIRT1 in the whole organism was not sufficiently potent to affect longevity although old mice presented lower levels of DNA damage, decreased expression of the ageing-associated and cell cycle regulator gene *p16Ink4a*[6], a better general health, and fewer spontaneous carcinomas and sarcomas. Pharmacological interventions directed to increase SIRT1 activity have been found to slow the onset of aging and delay age-associated disease. This concept has been validated in mice treated with the natural polyphenol resveratrol, which activates SIRT1 and produces changes associated with longer lifespan, including increased insulin

6 Note that *p16Ink4a* is frequently silenced in cancer through DNA methylation (see ▶ Chap. 8 of Santoro)

sensitivity, reduced insulin-like growth factor-1 (IGF-I) levels, increased AMP-activated protein kinase (AMPK) and peroxisome proliferator-activated receptor gamma coactivator 1-alpha (PGC-1α) activity, increased mitochondrial number, and improved motor function (Baur et al. 2006).

Like SIRT1, SIRT6 expression is also correlated with longevity. SIRT6 modulates telomeric chromatin through its association with telomeres and deacetylation of H3K9ac, which is required for the stable association of WRN, a factor that is mutated in the premature ageing disorder Werner syndrome. Accordingly, SIRT6-depleted cells show premature cellular senescence and exhibit abnormal telomere structures that resemble defects observed in Werner syndrome.

The seven members of sirtuin family are considered potential targets for the treatment of human pathologies including neurodegenerative diseases, cardiovascular diseases, and cancer. The action of sirtuins as epigenetic players in the regulation of fundamental biological pathways has prompted increased efforts to discover small molecules able to modify sirtuin activity.

9.3.2.2 PARPs

Another group of enzymes that uses NAD as a substrate and regulates cellular NAD levels are the PARP enzymes (◘ Fig. 9.3). There are 17 PARP-related enzymes. PARP1 and PARP2 catalyze the polymerization of ADP-ribose units from NAD, resulting in the attachment of either linear or branched poly-(ADP-ribose) (PAR) polymers to itself or other target proteins.

Many early studies on PARPs and ADP-ribosylation were carried out with PARP1, the most ubiquitous and abundant PARP family member. PARP1 has been implicated in epigenetic and chromatin regulation. PARP1 can modulate chromatin structure and act as a transcriptional coregulator. Initial evidence of a role of PARP1 in the regulation of chromatin structure derives from pioneering studies reporting that PARylation of nucleosomes causes chromatin decondensation *in vitro* (Poirier et al. 1982). In *Drosophila*, activation of PARP1 promotes decondensation of chromatin in response to heat shock or other cellular signaling pathways (Tulin and Spradling 2003). Accordingly, histone H1 and PARP1 exhibit a reciprocal pattern of chromatin binding at many RNA polymerase II transcribed promoters. PARP1 was enriched whereas H1 was depleted at these promoters. Furthermore, PARP1 can act as coregulator by altering the function of components of the transcriptional machinery.

PARP1 is strongly activated by DNA damage and, through the production of long PAR chains, leads to consumption of a large amount of cellular NAD. PARP1 and SIRT1 have a similar Km for NAD (PARP1 50–97 μM; SIRT1 94–96 μM). Consequentially, upon PARP1 activation, the decrease of NAD concentration should lead to a decrease in SIRT1 activity. PARP1 KO mice phenocopied many aspects of SIRT1 activity, such as a higher mitochondrial content, increased energy expenditure, and protection against metabolic diseases. The pharmacologic inhibition of PARP increases NAD content and SIRT1 activity and enhances oxidative metabolism. Interestingly, the NAD boosting effects of PARP inhibition enhance the activity of SIRT1 in the nucleus, but not that of SIRT2 in the cytoplasm or SIRT3 in mitochondria. This finding suggests that NAD is independently regulated within different cellular compartments. However, the role of PARP1 in aging and longevity is still not clear.

PARP inhibitors[7] are currently used in the management of tumors with absent or dysfunctional BRCA genes such as breast and ovarian cancers.

9.4 S-adenosylmethionine (SAM)

S-adenosylmethionine (SAM) is the cofactor of DNA and histone methyltransferases and is the second most common enzymatic cofactor after ATP (◘ Fig. 9.5).

9.4.1 Biosynthesis of SAM

SAM plays major roles in epigenetics, biosynthetic processes including phosphatidylcholine, creatine, and polyamine synthesis, as well as sulfur metabolism (Ducker and Rabinowitz 2017). The methyl group of SAM serves as donor for *trans*-methyltransferase reactions and is then converted into *S*-adenosylhomocysteine (SAH), which is a potent inhibitor of all methyltransferases (◘ Fig. 9.5).

The production of SAM and its control in the methylation of DNA and histones is a prominent example of how diet and nutrition can impact the epigenome. SAM is produced by one-carbon metabolism, which combines the folate cycle and the methionine cycle (◘ Fig. 9.6). One-carbon metabolism is an integrator of the nutrient status based on amino acids, glucose, and vitamins The products generated in these pathways are implicated in diverse cellular processes that include cellular biosynthesis, regulation of redox status, regulation of epigenetics through nucleic acid and protein methylation, and genome maintenance through the regulation of nucleotide pools.

Nutrients that fuel one-carbon metabolism are folate, serine, and glycine. Serine and glycine can be synthesized *de novo*. In contrast, folate can be synthesized by plants and most microorganisms whereas animals require dietary folate intake. In cells, folate is reduced into tetrahydrofolate (THF) by dihydrofolate reductase (DHFR). THF enters into the folate cycle and serves as scaffold that carries 1-carbon groups in a variety of reactions, including the production of methyl-THF (mTHF). The folate cycle is coupled to the methionine cycle through the demethylation of mTHF, which regenerate THF and allows methylation of homocysteine (hCYS) to produce methionine by methionine synthase and its cofactor vitamin B12. Methionine adenyltransferase (MAT) with ATP as cofactor converts methionine into SAM, which is then

---

7 The use of a PARP inhibitor in a BRCA-deficient cancers is the first example of the clinical application of the concept of synthetic lethality. BRCA1/2 genes produce proteins with an important role in repair of double strand DNA breaks (DSB) by homologous recombination (HR). While normal cells can repair DSBs through HR that is error free, loss of BRCA function forces cells to repair DSBs via non-homologous end joining NHEJ or the single-strand annealing sub-pathway of HR, both mechanisms being prone to error and genomic instability.

Ftnote_PARP1 and 2 are implicated in the repair of single-strand DNA break (SSB). PARP inhibitors trap the PARP complex and DNA replication stalls, converting SSBs into DSBs. When PARP inhibitors are given to cells with wild type BRCA1/2, the cells manage to repair DSBs by HR and survive. However, after administration of PARP inhibitors, tumor cells with mutated BRCA1/2 genes are unable to repair and undergo cell death.

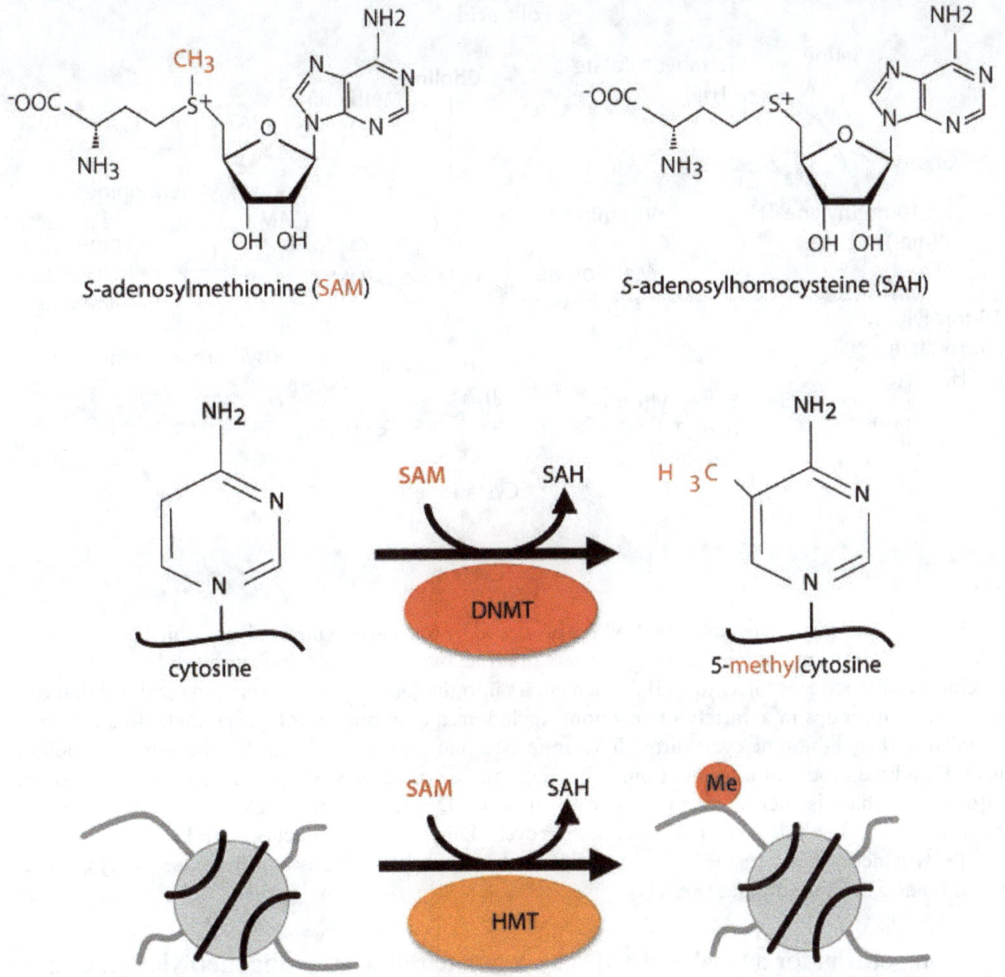

Fig. 9.5 *S*-adenosylmethionine (SAM) is the cofactor of DNA and histone methyltransferases. SAM and *S*-adenosylhomocystein (SAH) are shown. SAM is the cofactor for DNA methyltransferase and histone methyltransferase. The methyl group of SAM serves as donor for *trans*-methyltransferase reactions. SAM is then converted into SAH, which is a potent inhibitor of all methyltransferases

demethylated to form SAH. Deadenylation of SAH by *S*-adenosyl homocysteine hydrolase (SAHH) generates hCYS, resulting in a full turn of the methionine cycle.

9.4.2 SAM as Cofactor of DNA and Histone Methyltransferases

Enzymes that use SAM as cofactor are DNA methyltransferases (DNMTs), lysine methyltrasferases (KMTs), and peptidylarginine methyltransferases (PRMTs). The crosstalk between metabolism and chromatin is regulated by the kinetic properties of each enzyme and the physiological concentrations of the metabolites. This is particularly important for methyltransferases and acetyltransferases (see also Sect. 9.2.2), which respond to changes in metabolism due to low physiological concentrations of metabolic substrates that limit their enzymatic activities.

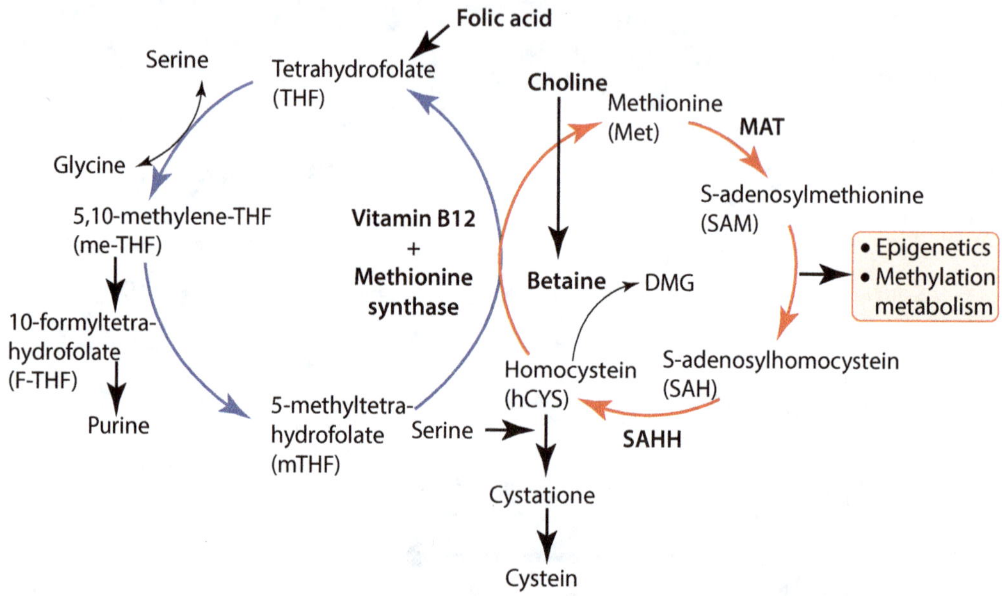

Fig. 9.6 One-carbon metabolism. SAM is produced by one-carbon metabolism, which combines the folate cycle and the methionine cycle. Nutrients that fuel one-carbon metabolism are folate, serine, and glycine. DHFR reduces folate to THF, which enters into the folate cycle and serves as scaffold that carries 1-carbon groups in a variety of reactions, including the production of mTHF. The folate cycle is coupled to the methionine cycle through Vitamin B12 that serves as cofactor for the demethylation of mTHF and the production of methionine by methionine synthase. MAT enzyme converts methionine into SAM, which is then demethylated to form SAH. Deadenylation of SAH by SAHH generates hCYS, resulting in a full turn of the methionine cycle. Dihydrofolate reductase (DHFR); tetrahydrofolate (THF); methyl-THF (mTHF); homocysteine (hCYS); methionine adenyltransferase (MAT); *S*-adenosyl homocysteine hydrolase (SAHH)

Competition for available SAM may regulate the contrasting methylation events, such as methylation of histone H3K4 that is associated with transcription activity and methylation of DNA and histone H3K9 that are associated with transcriptional silencing. Thus, changes in one-carbon metabolism, and, hence, alterations in the levels of SAM, can have an impact on gene expression. For instance, low amounts of external methyl donor groups from dietary sources can reduce the concentrations of SAM and affect gene expression. An important example of how nutrients affect the epigenome is the *agouti* mouse model where the maternal methyl dietary content affects the coat color of the offspring (Fig. 9.7 and 9.8). The murine *agouti* gene encodes a paracrine signaling molecule that signals follicular melanocytes to switch from producing black eumelanin to yellow phaeomelanin. The *agouti* (*a*) gene is transiently expressed in hair follicles resulting in a sub-apical yellow band on each black hair that causes the brown (agouti) coat color of wild-type mice. The *Agouti viable yellow* (A^{vy}) allele was first described in the early 1960s and resulted from the insertion of an intracisternal A particle (IAP) retrotransposon upstream of the transcription start site of the *agouti* gene. In mice carrying the A^{vy} allele, a cryptic promoter in the proximal end of the IAP promotes constitutive, ectopic *agouti* transcription, resulting in the expression not only in hair follicles but in all cells. This leads to a yellow fur, as well as adult-onset obesity, diabetes, and tumorigenesis (Fig.9.7). The activity of the A^{vy} allele is under the control of DNA methylation. In the case

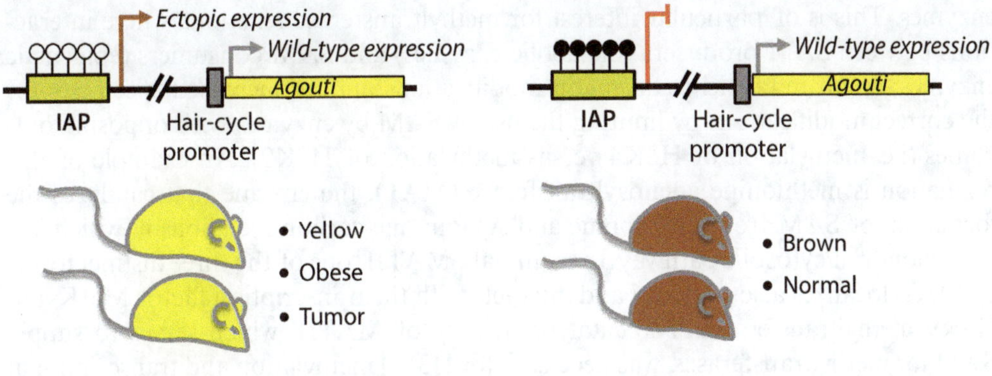

Fig. 9.7 Regulation of *Agouti viable yellow* (A^{vy}) by DNA methylation. The A^{vy} allele results from the insertion of an intracisternal A particle (IAP) retrotransposon upstream of the transcription start site of the *agouti* gene. DNA methylation at the IAP represses the ectopic expression of the *agouti* gene, resulting in healthy mice with brown coat color

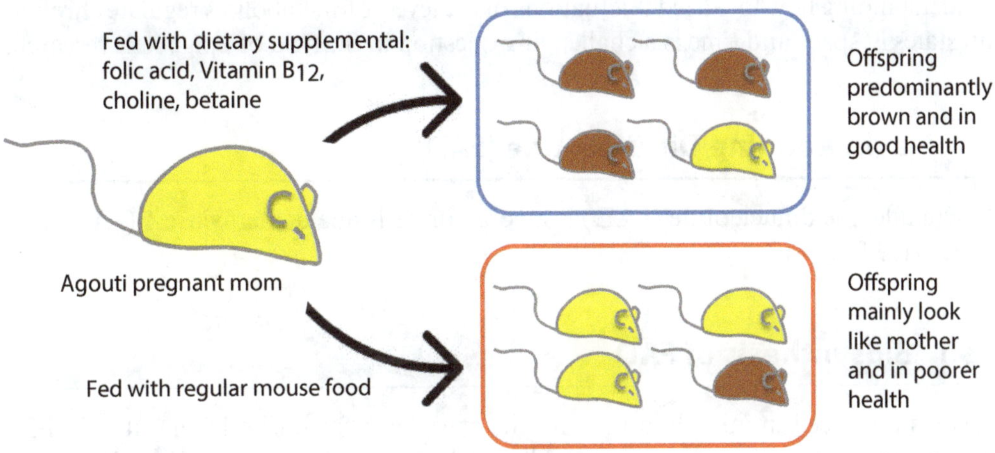

Fig. 9.8 Effect of maternal dietary supplementation on the phenotype and epigenotype of A^{vy} offspring. Maternal supplementation of A^{vy} mice with methyl donors (folate, betaine, vitamin B_{12}, and choline) results in litters with a higher proportion of offspring with brown coats, whereas the offspring of mothers fed with a regular diet have a predominantly yellow coat color

of hypermethylation, the allele is silent and produces a wild-type, agouti-colored coat (termed pseudoagouti), whereas hypomethylation characterizes a transcriptionally active state that produces a completely yellow coat and increases the tendency towards obesity and cancer (Morgan et al. 1999) (Fig. 9.7).

Thus, agouti and pseudoagouti mice are genetically identical but they differ in epigenetic state and phenotype. Remarkably, if a pregnant A^{vy} mouse receives dietary supplements such as vitamin B12, folic acid, choline, and betaine, fueling the one-carbon metabolism and leading to an increase in SAM production, the IAP in offspring will be methylated and *agouti* shows wild-type expression (Fig. 9.8). These pups have a brown fur and no increased tendency towards obesity and cancer. Thus, environmental exposure or nutritional status can alter gene expression, especially during embryonic development, when the epigenome is established.

Chromatin-localized biosynthesis of metabolites has emerged as a process that modulates the expression of nearby genes by providing metabolites for epigenetic enzymes. This is of particular interest for methyltransferase reactions where interactions between SAM producers (metabolic enzymes) and SAM consumers (epigenetic enzymes) result in efficient chromatin modifications and an increased specificity for the correct modifications by limiting the use of SAM by enzymes with opposing outcomes (i.e. methylation of H3K4 versus methylation of H3K9). An example of this regulation is methionine adenosyltransferase (MAT), the enzyme that catalyzes the formation of SAM from methionine and ATP in one-carbon metabolism, which is a mitochondrial/cytosolic pathway. In mammals, MATII, one of the three distinct forms of MAT, localizes also to nuclei and interacts with the transcription factor MafK and H3K9 methyltransferases. The catalytic activity of MATII, which serves to supply SAM for methyltransferases, was necessary for H3K9 methylation and transcriptional repression of the *heme oxygenase-1* gene (Katoh et al. 2011). The presence of microdomains for chromatin modifications, where SAM moieties for histone or DNA methyltransferases are immediately replenished after the modification, was also described for other epigenetic enzymes using, e.g., acetyl-CoA and NAD (see ▶ Sects. 9.2 and 9.3). Understanding how localized fluctuations in the levels of metabolites regulate chromatin states in space and time is a challenging question in the field of epigenetic research.

9.5 Flavin Adenine Dinucleotide (FAD)

Flavin adenine dinucleotide (FAD) is a cofactor of lysine demethylase 1 (KDM1 or LSD1) (◘ Fig. 9.9).

9.5.1 Biosynthesis of FAD

FAD is produced in mitochondria and the cytoplasm from riboflavin (vitamin B2), which can be found in many vegetables and meat. Riboflavin is phosphorylated by riboflavin kinase and ATP to generate riboflavin 5-phosphate, known as flavin

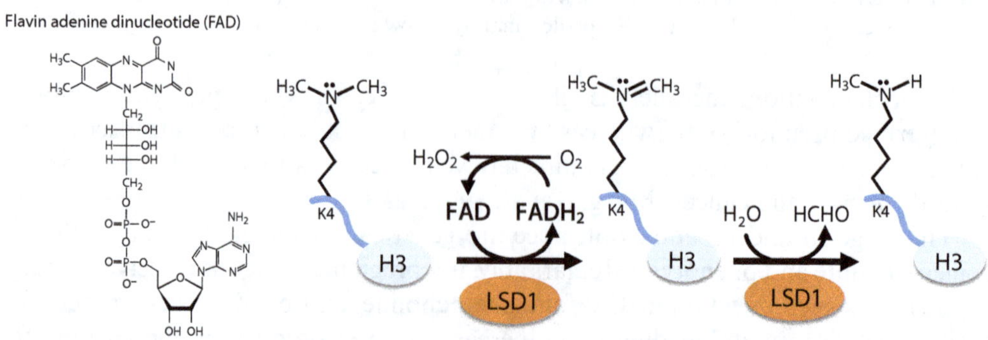

◘ **Fig. 9.9** Flavin adenine dinucleotide (FAD) is a cofactor of lysine demethylase 1 (LSD1). The structure of FAD and its involvement in the histone demethylation mediated by LSD1 are shown. The reaction mediated by LSD1 requires protonation of the nitrogen to be demethylated, which limits LSD1 demethylating abilities to mono- and dimethylated lysines. LSD1-mediated demethylation generates formaldehyde and FADH2. Recycling of FAD occurs via oxidation of FADH2, which produces peroxide

monucleotide (FMN), which is further converted to FAD by FMN adenyltransferase and ATP. Enzymes involved in FAD synthesis (FAD synthase) and hydrolysis (FAD pyrophosphatase) can also be found in the nucleus, providing a dynamic pool of nuclear FAD by balancing the ratio between its oxidized (FAD^+) and reduced (FADH2) forms. The reduction of FAD to FADH2 is an essential reaction in the TCA cycle, where succinate dehydrogenase requires covalently bound FAD to catalyze the oxidation of succinate to fumarate. Finally, FADH2 can be re-oxidized by oxygen producing FAD and peroxide.

FAD acts as cofactor of flavoenzymes, which have dehydrogenase, oxidase, monooxygenase, or reductase activities. Flavoenzymes play crucial roles in bioenergetics, photochemistry, bioluminescence, redox homeostasis, chromatin remodeling, DNA repair, protein folding, apoptosis, and other physiologically relevant processes (Joosten and van Berkel 2007). Deficiency in FAD-dependent enzymes and/or impairment of flavin homeostasis in humans and animal models have been linked to several diseases, including cancer and neurological disorders.

9.5.2 FAD as Cofactor of Lysine Demethylase 1 (LSD1)

The majority of reported flavoenzymes localize to the mitochondria or cytoplasm. LSD1 is one of a few flavoproteins in the nucleus and is the only epigenetic enzyme that utilizes FAD as an essential cofactor for catalytic activity. LSD1 is a member of the flavin-containing amine oxidase family and represses transcription by removing the methyl group from mono-methylated and di-methylated H3K4 (Shi et al. 2004) (◘ Fig. 9.9). LSD1, in association with a nuclear receptor, is also involved in the demethylation of H3K9. Furthermore, demethylation by LSD1 was also reported for non-histone proteins such as p53. FAD-dependent monoamine oxidases typically catalyze the oxidation of amine-containing substrates, utilizing molecular oxygen as electron acceptor. The reaction mediated by LSD1 requires protonation of the nitrogen to be demethylated, which limits LSD1 demethylating abilities to mono- and dimethylated lysines. Of note is that LSD1-mediated demethylation generates formaldehyde and FADH2. Moreover, recycling of FAD occurs via oxidation of FADH2, which produces peroxide. The generation of formaldehyde and peroxide might potentially have deleterious effects when present near promoters. However, how cells deal with these products remains yet to be investigated.

The important function of LSD1 is evident by the embryonic lethality in null-mutant *Kdm1*^{KO} mice and LSD1-deficient ESCs displaying global DNA hypomethylation (Wang et al. 2009). The biological significance of FAD-dependent LSD1 activities in metabolic regulation have recently been supported by studies showing a crosstalk between energy metabolism and epigenetic gene regulation, e.g., by LSD1 repressing energy-expenditure genes such as *PGC1*α through H3K4 demethylation in adipocytes where excess energy is stored as triglycerides. Cellular FAD content remarkably increased during adipogenic differentiation while a reduction in FAD levels by downregulation of riboflavin kinase induced expression of most LSD1-target genes. Moreover, transgenic overexpression of LSD1 in adipose tissue leads

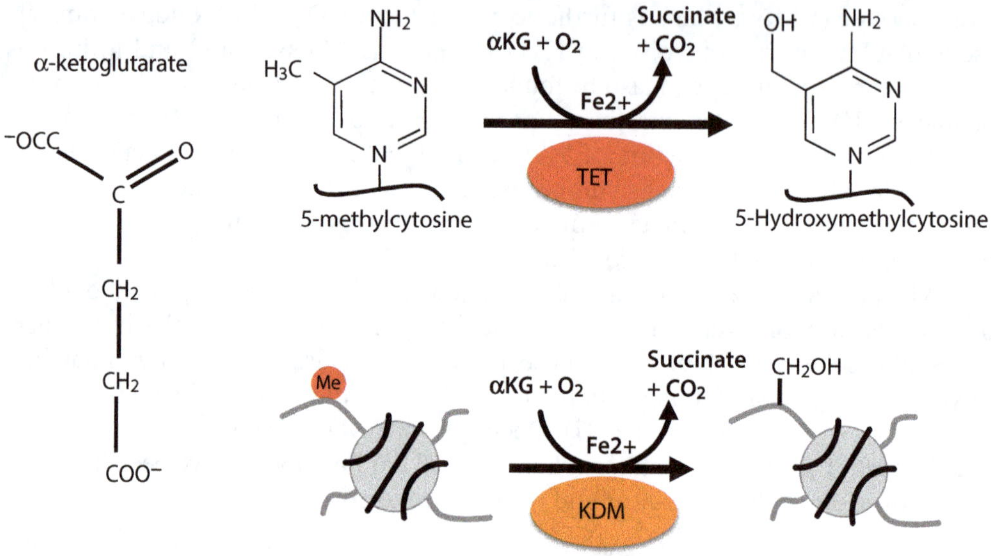

Fig. 9.10 α-ketoglutarate (αKG) is a co-substrate for dioxygenases. αKG is a required cosubstrate for Jumonji C-family histone demethylases (KDMs) and TET enzymes, which participate in a multi-step DNA demethylation process. TETs and KDMs require Fe(II) as a cofactor metal and α-KG as a cosubstrate to catalyze the reactions in which one oxygen atom from molecular oxygen (O_2) is attached to a hydroxyl group in the substrate (hydroxylation) while the other is taken up by αKG, leading to the decarboxylation of αKG and subsequent release of carbon dioxide (CO_2) and succinate

to increased browning[8] by promoting mitochondrial activity, thus protecting against diet-induced obesity. Finally, LSD1 inhibitors lead to a shift from oxidative to glycolytic metabolism in brown adipose tissue, resulting in weight gain but improved glucose tolerance (Duteil et al. 2016). LSD1 inhibitors have entered in clinical trials to treat mixed lineage leukemia.

9.6 α-Ketoglutarate (αKG)

α-ketoglutarate (αKG) is an intermediate of the TCA cycle and is a required cosubstrate for Jumonji domain-containing histone lysine demethylases (KDMs) and TET enzymes, which participate in a multi-step DNA demethylation process (Fig.9.10).

9.6.1 Biosynthesis of α-Ketoglutarate

Isocitrate dehydrogenases (IDH1, IDH2, IDH3) and glutamate dehydrogenase (GDH) are the enzymes producing αKG (Fig. 9.11).

IDH1 and IDH2 are nicotinamide adenine dinucleotide phosphate ($NADP^+$)-dependent enzymes whereas IDH3 is a NAD-dependent enzyme. IDH1 is localized to the cytoplasm and peroxisomes. IDH2 and IDH3 are localized to the mitochon-

8 Brown adipose tissue that has the function to dissipates energy to produce heat and has the potential to regulate body temperature by thermogenesis.

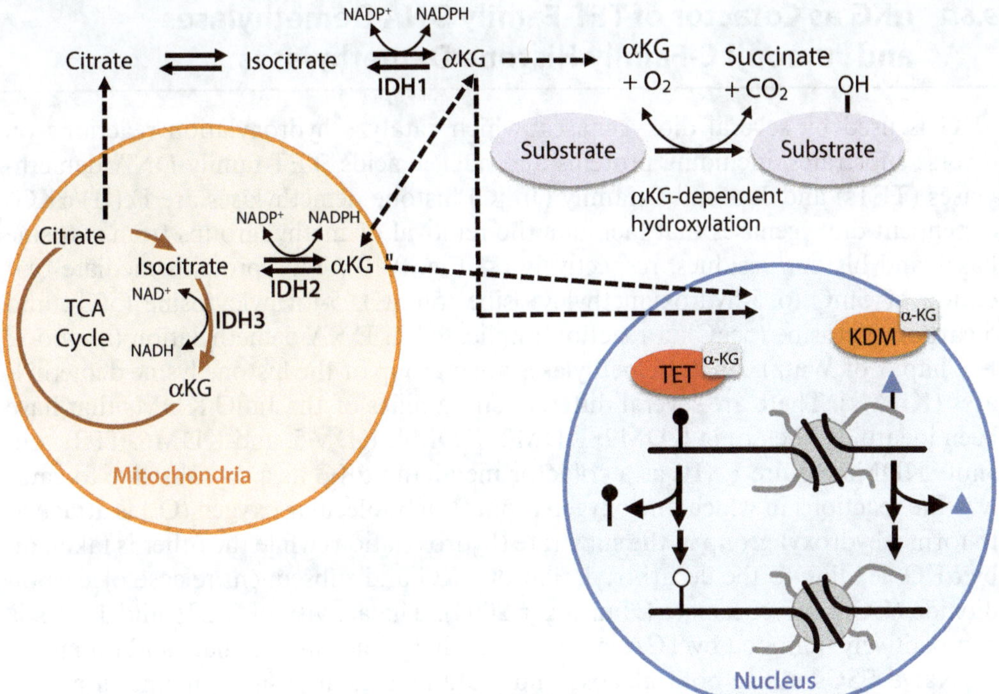

Fig. 9.11 Biosynthesis of α-ketoglutarate. Isocitrate dehydrogenases (IDH1, IDH2, IDH3) and glutamate dehydrogenase (GDH) are the enzymes producing αKG from isocitrate. IDH1 is localized to the cytoplasm whereas IDH2 and IDH3 are localized to the mitochondria. IDH1 and IDH2 are $NADP^+$-dependent enzymes whereas IDH3 is a NAD-dependent enzyme

dria and play a key role in the regulation of the TCA cycle. IDH1 and IDH2 are bidirectional enzymes that can both produce and consume αKG to meet cellular demands. αKG is used for four separate pathways: TCA cycle, anaplerosis[9], fatty acid synthesis, and protein and nucleic acid hydroxylation. αKG is a rate-determining intermediate in the TCA cycle and has a crucial role in cellular energy metabolism. In the TCA cycle, αKG is decarboxylated to succinyl-CoA and CO_2 by the oxoglutarate dehydrogenase complex (OGDC), a key control point of the TCA cycle. αKG also plays a major role in anaplerosis of the TCA cycle, a pathway that replenishes TCA intermediates from other biosynthetic pathways. In this case, αKG serves as entry point into TCA cycle for several 5-carbon amino acids (Arg, Glu, Gln, His, and Pro) that, after the conversion into glutamate, undergo oxidative deamination catalyzed by GDH to form αKG and NH3. αKG can also be reduced by IDH1 and IDH2 to isocitrate and then citrate. As discussed below, αKG is used as a cosubstrate for multiple αKG-dependent dioxygenases that are involved in the hydroxylation of various proteins and nucleic acids.

9 Anaplerosis refers to metabolic pathways that replenish the TCA cycle intermediates.

9.6.2 αKG as Cofactor of TET-Family DNA Demethylases and Jumonji C-Family Histone Demethylases

αKG is used by several dioxygenases, which catalyze hydroxylation reactions on diverse substrates, including proteins and nucleic acids. TET-family DNA demethylases (TETs) and Jumonji C-family (JmjC) histone demethylases are Fe(II)/αKG-dependent dioxygenases that facilitate the removal of methyl groups from cytosine bases and histone residues, respectively (◘ Fig. 9.10). TET proteins mediate oxidation of 5mC to 5-hydroxymethylcytosine (5hmC), 5-formylcytosine (5fC), and 5-carboxylcytosine (5caC), a reaction implicated in DNA demethylation (see book ► Chap. 1 of Wutz). JmJC demethylases are a group of the histone lysine demethylases (KDMs). There are several different subfamilies of the JmjC KDMs that have been identified, including KDM2, KDM3, KDM4, KDM5, and KDM6. TETs and JmjC KDMs require Fe(II) as a cofactor metal and αKG as a co-substrate to catalyze the reactions in which one oxygen atom from molecular oxygen (O_2) is attached to form a hydroxyl group in the substrate (hydroxylation) while the other is taken up by αKG, leading to the decarboxylation of αKG and subsequent release of carbon dioxide (CO_2) and succinate (Hausinger 2004). The activity of TETs and JmjCs is competitively inhibited by TCA cycle intermediates such as succinate and fumarate.

Naive ESCs utilize both glucose and glutamine catabolism to maintain a high level of αKG. The elevated αKG to succinate ratio in naive ESCs promotes histone and DNA demethylation and maintains pluripotency. Direct manipulation of the intracellular αKG/succinate ratio is sufficient to regulate multiple chromatin modifications, including H3K27me3 and TET-dependent DNA demethylation, which contribute to the regulation of pluripotency-associated gene expression.

Initial genome-wide sequencing studies in patients with glioblastoma identified somatic mutations in *IDH1* and tumors without *IDH1* mutations often have mutations affecting *IDH2* (◘ Fig. 9.12) (Parsons et al. 2008; Yan et al. 2009). Later, a similar mutation was also found in acute myelogenous leukemia (AML) and chondrosarcomas. Mutations in *IDH1* and *IDH2* are almost always in the binding site for isocitrate (Arg 132 for *IDH1*, Arg 172 for *IDH2*). However, mutant *IDH1* and *IDH2* are not simply inactive enzymes but, instead, possess a novel enzymatic activity that leads to the production of 2-hydroxyglutarate (2HG) (Dang et al. 2009). Accordingly, human malignant gliomas and AMLs harboring *IDH1* mutations contained elevated levels of 2HG.

Evidence indicates that 2HG may act as an oncometabolite by competitive inhibition of αKG-dependent enzymes, including both JmjC KDMs and TETs (◘ Fig. 9.12). The structure of 2HG and αKG is similar with the exception of the 2-ketone group in αKG is replaced by a hydroxyl group in 2HG. Structural analysis showed that 2HG occupies the same space as αKG does in the active site of histone demethylases and adopted a nearly identical orientation as αKG, thereby preventing the binding of αKG to the active site of the enzymes.

In vitro studies revealed that 2HG strongly inhibits histone demethylases. The strongest inhibition was observed with KDM4A, which demethylates H3K9 and H3K36, followed by the H3K9/H3K36 demethylase KDM4C and the H3K36 demethylase KDM2A. These finding were supported by *in vivo* studies showing that human tumors expressing *IDH1* and *IDH2* mutants had increased histone methylation at H3K4, H3K9, H3K27, and H3K79 (Lu et al. 2012).

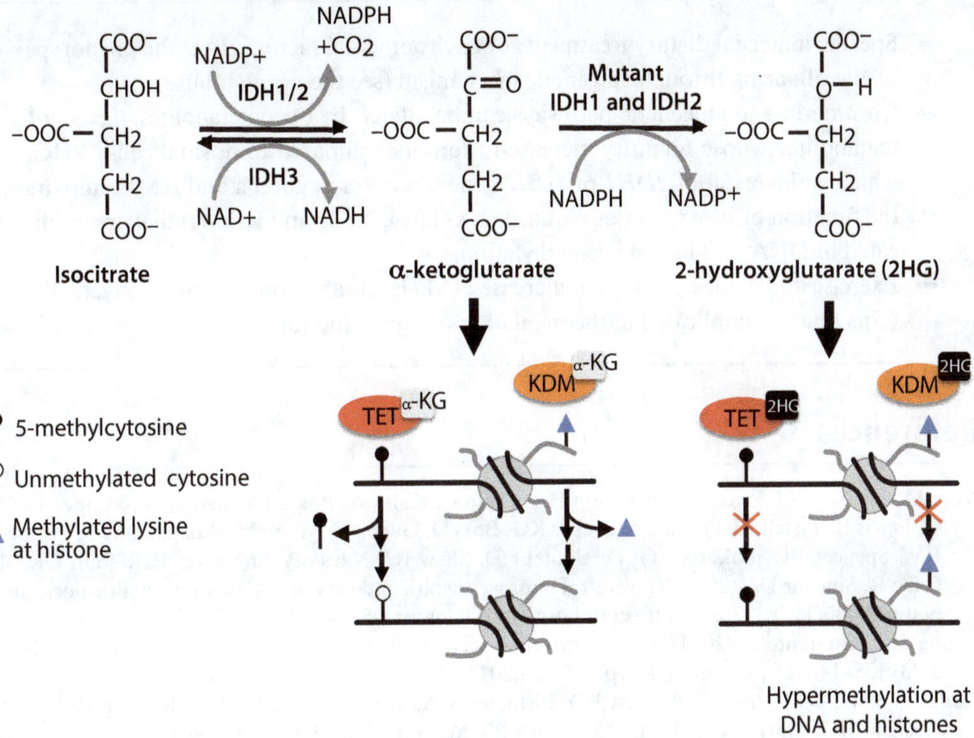

Fig. 9.12 Chemical reactions catalyzed by IDH enzymes and IDH1/2 mutants. *IDH1* and *IDH2* mutants possess a novel enzymatic activity that leads to the production of 2-hydroxyglutarate (2HG). 2HG acts as competitive inhibitor of αKG-dependent enzymes, including both KDMs and TETs, thereby causing hypermethylation at DNA and histones

The second major target of *IDH1* and *IDH2* mutations is the TET family. 2-HG inhibits TET activity and this inhibition could be overcome by the addition of αKG (Xu et al. 2011). Furthermore, glioblastomas expressing mutant *IDH1* display hypermethylation at a large number of loci, leading to what is known as the glioma-CpG island methylator phenotype. Remarkably, mutations affecting *TET2* and *IDH1* in AML (see ▶ Chap. 8 of Santoro) occur in a mutually exclusive manner, suggesting that their biological effect is similar and that they have overlapping roles in AML pathogenesis.

Take-Home Message

- Most chromatin-modifying enzymes use metabolites that are central to cellular metabolism as cofactor
- The epigenome acts a sensor of the whole metabolic network and regulates gene expression accordingly.
- Environmental and physiological stimuli can alter metabolite concentration, which in turn might affect gene expression by modulating the activity of epigenetic and chromatin enzymes.
- Chromatin-localized biosynthesis of metabolites has emerged as a process that modulates the expression of nearby genes by providing metabolites that, as cofactors, activate epigenetic enzymes.

- Specific maternal dietary treatments or environmental factors affect the phenotype of the offspring through epigenetic mechanism (see the *agouti* mouse).
- Chromatin and epigenetic pathways can be altered by oncometabolites, a class of metabolites whose quantity increase in tumors compared to normal cells. 2-HG, which is the result of *IDH1* or *IDH2* gene mutation in glioma and AML, impairs the function of dioxygenases requiring αKG (e.g. TETs and KDMs) that are implicated in DNA and histone demethylation.
- Exercise and caloric restriction increase NAD levels and consequently activate Sirtuins that are implicated in the regulation of aging and longevity.

References

Baur JA, Pearson KJ, Price NL, Jamieson HA, Lerin C, Kalra A, Prabhu VV, Allard JS, Lopez-Lluch G, Lewis K, Pistell PJ, Poosala S, Becker KG, Boss O, Gwinn D, Wang M, Ramaswamy S, Fishbein KW, Spencer RG, Lakatta EG, Le Couteur D, Shaw RJ, Navas P, Puigserver P, Ingram DK, de Cabo R, Sinclair DA (2006) Resveratrol improves health and survival of mice on a high-calorie diet. Nature 444(7117):337–342. https://doi.org/10.1038/nature05354

Chang HC, Guarente L (2014) SIRT1 and other sirtuins in metabolism. Trends Endocrinol Metab 25(3):138–145. https://doi.org/10.1016/j.tem.2013.12.001

Dang L, White DW, Gross S, Bennett BD, Bittinger MA, Driggers EM, Fantin VR, Jang HG, Jin S, Keenan MC, Marks KM, Prins RM, Ward PS, Yen KE, Liau LM, Rabinowitz JD, Cantley LC, Thompson CB, Vander Heiden MG, Su SM (2009) Cancer-associated IDH1 mutations produce 2-hydroxyglutarate. Nature 462(7274):739–744. https://doi.org/10.1038/nature08617

Ducker GS, Rabinowitz JD (2017) One-carbon metabolism in health and disease. Cell Metab 25(1):27–42. https://doi.org/10.1016/j.cmet.2016.08.009

Duteil D, Tosic M, Lausecker F, Nenseth HZ, Muller JM, Urban S, Willmann D, Petroll K, Messaddeq N, Arrigoni L, Manke T, Kornfeld JW, Bruning JC, Zagoriy V, Meret M, Dengjel J, Kanouni T, Schule R (2016) Lsd1 ablation triggers metabolic reprogramming of brown adipose tissue. Cell Rep 17(4):1008–1021. https://doi.org/10.1016/j.celrep.2016.09.053

Frederick DW, Loro E, Liu L, Davila A Jr, Chellappa K, Silverman IM, Quinn WJ 3rd, Gosai SJ, Tichy ED, Davis JG, Mourkioti F, Gregory BD, Dellinger RW, Redpath P, Migaud ME, Nakamaru-Ogiso E, Rabinowitz JD, Khurana TS, Baur JA (2016) Loss of NAD homeostasis leads to progressive and reversible degeneration of skeletal muscle. Cell Metab 24(2):269–282. https://doi.org/10.1016/j.cmet.2016.07.005

Hausinger RP (2004) FeII/alpha-ketoglutarate-dependent hydroxylases and related enzymes. Crit Rev Biochem Mol Biol 39(1):21–68. https://doi.org/10.1080/10409230490440541

Joosten V, van Berkel WJ (2007) Flavoenzymes. Curr Opin Chem Biol 11(2):195–202. https://doi.org/10.1016/j.cbpa.2007.01.010

Kaelin WG Jr, McKnight SL (2013) Influence of metabolism on epigenetics and disease. Cell 153(1):56–69. https://doi.org/10.1016/j.cell.2013.03.004

Katoh Y, Ikura T, Hoshikawa Y, Tashiro S, Ito T, Ohta M, Kera Y, Noda T, Igarashi K (2011) Methionine adenosyltransferase II serves as a transcriptional corepressor of Maf oncoprotein. Mol Cell 41(5):554–566. https://doi.org/10.1016/j.molcel.2011.02.018

Lin SJ, Defossez PA, Guarente L (2000) Requirement of NAD and SIR2 for life-span extension by calorie restriction in Saccharomyces cerevisiae. Science 289(5487):2126–2128

Lu C, Ward PS, Kapoor GS, Rohle D, Turcan S, Abdel-Wahab O, Edwards CR, Khanin R, Figueroa ME, Melnick A, Wellen KE, O'Rourke DM, Berger SL, Chan TA, Levine RL, Mellinghoff IK, Thompson CB (2012) IDH mutation impairs histone demethylation and results in a block to cell differentiation. Nature 483(7390):474–478. https://doi.org/10.1038/nature10860

Mills KF, Yoshida S, Stein LR, Grozio A, Kubota S, Sasaki Y, Redpath P, Migaud ME, Apte RS, Uchida K, Yoshino J, Imai SI (2016) Long-term administration of nicotinamide mononucleotide mitigates age-associated physiological decline in mice. Cell Metab 24(6):795–806. https://doi.org/10.1016/j.cmet.2016.09.013

Morgan HD, Sutherland HG, Martin DI, Whitelaw E (1999) Epigenetic inheritance at the agouti locus in the mouse. Nat Genet 23(3):314–318. https://doi.org/10.1038/15490

Moussaieff A, Rouleau M, Kitsberg D, Cohen M, Levy G, Barasch D, Nemirovski A, Shen-Orr S, Laevsky I, Amit M, Bomze D, Elena-Herrmann B, Scherf T, Nissim-Rafinia M, Kempa S, Itskovitz-Eldor J, Meshorer E, Aberdam D, Nahmias Y (2015) Glycolysis-mediated changes in acetyl-CoA and histone acetylation control the early differentiation of embryonic stem cells. Cell Metab 21(3):392–402. https://doi.org/10.1016/j.cmet.2015.02.002

Parsons DW, Jones S, Zhang X, Lin JC, Leary RJ, Angenendt P, Mankoo P, Carter H, Siu IM, Gallia GL, Olivi A, McLendon R, Rasheed BA, Keir S, Nikolskaya T, Nikolsky Y, Busam DA, Tekleab H, Diaz LA Jr, Hartigan J, Smith DR, Strausberg RL, Marie SK, Shinjo SM, Yan H, Riggins GJ, Bigner DD, Karchin R, Papadopoulos N, Parmigiani G, Vogelstein B, Velculescu VE, Kinzler KW (2008) An integrated genomic analysis of human glioblastoma multiforme. Science 321(5897):1807–1812. https://doi.org/10.1126/science.1164382

Poirier GG, de Murcia G, Jongstra-Bilen J, Niedergang C, Mandel P (1982) Poly(ADP-ribosyl)ation of polynucleosomes causes relaxation of chromatin structure. Proc Natl Acad Sci U S A 79(11):3423–3427

Shi Y, Lan F, Matson C, Mulligan P, Whetstine JR, Cole PA, Casero RA (2004) Histone demethylation mediated by the nuclear amine oxidase homolog LSD1. Cell 119(7):941–953. https://doi.org/10.1016/j.cell.2004.12.012

Tulin A, Spradling A (2003) Chromatin loosening by poly(ADP)-ribose polymerase (PARP) at Drosophila puff loci. Science 299(5606):560–562. https://doi.org/10.1126/science.1078764. 299/5606/560 [pii]

Verdin E (2015) NAD(+) in aging, metabolism, and neurodegeneration. Science 350(6265):1208–1213. https://doi.org/10.1126/science.aac4854

Wang J, Hevi S, Kurash JK, Lei H, Gay F, Bajko J, Su H, Sun W, Chang H, Xu G, Gaudet F, Li E, Chen T (2009) The lysine demethylase LSD1 (KDM1) is required for maintenance of global DNA methylation. Nat Genet 41(1):125–129. https://doi.org/10.1038/ng.268

Wellen KE, Hatzivassiliou G, Sachdeva UM, Bui TV, Cross JR, Thompson CB (2009) ATP-citrate lyase links cellular metabolism to histone acetylation. Science 324(5930):1076–1080. https://doi.org/10.1126/science.1164097

Xu W, Yang H, Liu Y, Yang Y, Wang P, Kim SH, Ito S, Yang C, Xiao MT, Liu LX, Jiang WQ, Liu J, Zhang JY, Wang B, Frye S, Zhang Y, Xu YH, Lei QY, Guan KL, Zhao SM, Xiong Y (2011) Oncometabolite 2-hydroxyglutarate is a competitive inhibitor of alpha-ketoglutarate-dependent dioxygenases. Cancer Cell 19(1):17–30. https://doi.org/10.1016/j.ccr.2010.12.014

Yan H, Parsons DW, Jin G, McLendon R, Rasheed BA, Yuan W, Kos I, Batinic-Haberle I, Jones S, Riggins GJ, Friedman H, Friedman A, Reardon D, Herndon J, Kinzler KW, Velculescu VE, Vogelstein B, Bigner DD (2009) IDH1 and IDH2 mutations in gliomas. N Engl J Med 360(8):765–773. https://doi.org/10.1056/NEJMoa0808710

Zhang H, Ryu D, Wu Y, Gariani K, Wang X, Luan P, D'Amico D, Ropelle ER, Lutolf MP, Aebersold R, Schoonjans K, Menzies KJ, Auwerx J (2016) NAD(+) repletion improves mitochondrial and stem cell function and enhances life span in mice. Science 352(6292):1436–1443. https://doi.org/10.1126/science.aaf2693

Glossary

A/B compartment On a large scale, chromosomes are organized into two distinct compartments A and B, whereby A refers to active and B to inactive regions with reference to gene expression. A/B compartments were observed in chromatin conformation capture (Hi-C) studies. Researchers observed that contacts of DNA sequences of individual chromosomes indicated two spatial compartments. Regions in compartment A interacted preferentially with A compartment-associated regions. Similarly, interactions within B compartments were observed but not between A and B compartments. A/B compartment-associated regions are several mega base pairs long. Whereas A compartments contain active genes, B compartments are considered largely inactive.

Aleurone The aleurone is the outermost cell layer of the endosperm in cereals, which surrounds the inner, starchy endosperm. In maize, the aleurone produces anthocyanin pigments that lead to the purple-brownish color of wild-type kernels.

Allelic/biallelic/monoallelic expression A diploid genome comprises two copies of each chromosome. A gene can be expressed from both chromosomes (biallelic expression), or from only one (monoallelic expression). In case of genomic imprinting a gene is expressed only from either the maternally or the paternally inherited chromosome. An analysis that takes into account from which allele transcripts originate is referred to an analysis of allelic expression.

Anabolism The process by which the body utilizes energy to synthesize complex molecules.

Anaplerosis Metabolic pathways that replenish the citric acid (TCA) cycle intermediates, which are essential to energy metabolism.

Angiogenesis The process of developing new blood vessels. Angiogenesis is critically important during the normal development of the embryo and fetus. It also appears to be important during tumor formation.

Anther The anther is the male reproductive organ of flowering plants and the site of meiosis and male gametogenesis. After meiosis, all four meiotic products, the microspores, survive to form multicellular pollen grain through two mitotic divisions. The pollen contains two sperm cells that participate in double fertilization.

Bivalent promoter Bivalent promoters are characterized by both repressing and activating epigenetic regulators, which is most commonly observed as dual-modified chromatin with H3K27me3 and H3K4me3 in the same region. Associated genes are expressed at low levels and often developmentally regulated. Bivalent chromatin is observed in stem cells or progenitor cells and is resolved to either repressed and H3K27me3 marked chromatin or active and H3K4me3 marked chromatin, when these cells differentiate into different lineages.

Blastema A blastema is a mass of proliferating, undifferentiated cells that has the capability to regenerate organs and tissues. Cells appear not to be pluripotent, but to retain the restricted developmental potential of the tissue of origin.

Blastocyst The blastocyst represents an early mammalian embryonic developmental stage. It consists of the inner cell mass (ICM) which upon further development contribute to the formation of the embryo. The outer layer of the blastocyst is called the trophoblast which gives rise to the placenta. A cavity inside, termed blastocoel, is filled with fluid. The cells of the ICM are pluripotent and can be maintained in culture as embryonic stem cells.

Bromodomain The bromodomain was identified by sequence homology in the Drosophila gene *Brahma*, from which its name is derived, and genes involved in transcriptional activation. Bromodomains are approximately 110 amino acids in length and recognize acetylated lysines in the N-terminal tails of histones. Among bromodomain-containing proteins are histone acetyltransferases and the histone methyltransferase ASH1L.

Central cell The central cell is the second female gamete of flowering plants that participates in double fertilization. In most species, it is genetically identical to the egg cell as both female gametes are derived from the same meiotic product (functional megaspore) through mitotic divisions. However, in the majority of cases, it carries two copies of the maternal genome and is, thus, homo-diploid as compared to the haploid egg cell.

Chemoresistance The resistance of a tumor to chemotherapy.

Chromodomain The chromodomain is defined by sequence homology and is found as domain of about 40–50 amino acids in proteins associated with chromatin. Chromodomain containing proteins include binders of methylated histones. The chromodomain of HP1 has affinity for H3K9me3, and the chromodomain of Polycomb has affinity for H3K27me3.

Chromosome territory A chromosome territory refers to the volume that a particular chromosome occupies in the cell nucleus. Chromosome territories have a globular shape and a diameter of a few micrometers. The chromosomal DNA in interphase adopts a loosely folded configuration. Although defined borders between chromosome territories have not been observed, chromosomes occupy discrete regions.

CIMP CpG island methylator phenotype defined as frequent methylation of multiple CpG islands that were found in many types of cancer.

Circadian clocks Endogenous oscillators that control 24-hour physiological and behavioral processes in organisms. These cell-autonomous clocks are composed of a transcription/translation-based autoregulatory feedback loop.

CpG island They are short interspersed DNA sequences (500–1500 bp long) with elevated CG density compared to the majority of the genome. Approximately 70%

of annotated gene promoters in the human and mouse genome are associated with a CGI, making this the most common promoter type in the vertebrate genome. CpG islands are normally devoid of any methylation at cytosines. The “p” represents the phosphor-ribose linkage and serves to clarify that CpG islands refer to CG dinucleotides and not GC.

Crypt In general, crypts are anatomical structures that are narrow but deep invaginations into a larger structure. In this context, the intestinal epithelium is organized in a basal crypt and a villus. Crypts of the small intestine contain replicating stem cells, Paneth cells of the innate immune system and goblet cell producing mucus. After stem cell division, one of the two daughter cells remain in the crypt as a stem cell, while the other differentiates and migrates up the side of the crypt and eventually forms specifically differentiated cells in the villus.

Cuticular structure Insects are protected by a rigid exoskeleton made out of chitin. The segmental structure of the insect body is particularly well reflected in the organization the thorax and abdomen.

Double fertilization Double fertilization is typical of flowering plants, where reproduction involves two fertilization events: one sperm cell fuses with the egg cell to form the zygote while a second sperm fuses with the central cell that develops into the endosperm. These two fertilization products develop coordinately into a seed, surrounded by maternal tissues that form the seed coat. In the majority of species, each pair of male and female gametes are genetically identical, respectively, i.e., they are derived from the same meiotic product.

Genomic repeat Genomic repeats are DNA sequences that occur in multiple copies throughout the genome. Repeats can be subdivided into tandem repeats and interspersed repeats, the latter reflecting different types of transposable elements. Tandem repeats are adjacent to each other in the same or opposite orientation, e.g., satellite repeats typical of centromeres, minisatellites (10–60 bp) found in many places in the genome including centromeres, and microsatellites (<10 bp) found throughout the genome and typical of telomeres.

Glycolysis Metabolic pathway that breaks down glucose into two three-carbon compounds and generates energy through the synthesis of ATP and NADH.

Hemolymph Insects have an open circulatory system. The hemolymph, like the blood system in vertebrates, among other tasks is responsible for carrying nutrients to organs and transport of cells of the innate immune system to damaged or infected parts.

Heterochromatin The two forms of chromatin were distinguished by cytological staining of cells. Heterochromatin stained darker than euchromatin due to a denser packing and is generally localized to the periphery of the nucleus. However, it has become clear that there are distinct forms of heterochromatin that can be molecularly distinguished. Constitutive heterochromatin at the pericentric repeats is characterized by H3K9me3 and HP1 binding, whereas facultative heterochromatin of the inactive X chromosome in female mammals is characterized by H3K27me3 and PcG binding.

Histolyzed In flies like Drosophila, during embryogenesis cell lineages forming the adult structures (like the imaginal discs) are developed separately from the larval organs and tissues. During metamorphosis the molting hormone ecdysone induces the formation of a puparium. During this period the imaginal structures form the organ and appendages of the adult fly. Conversely the larval structures are decaying (they become histolyzed).

Histone code The histone code has been a hypothesis stating that chemical modifications to histone proteins would regulate the transcription of genetic information in a combinatorial manner. Together with DNA methylation this concept has been extended to an epigenetic code. Although such a code is far from the clarity and completeness of the codons for amino acids that are used by the ribosome for translation, the concepts still remains useful for discussing combinatorial interactions between histone modifications and their effect on gene regulation. The present view has abandoned with the generality of the histone code and adopted a biochemical and molecular view of processes that work on chromatin and involve histone modifications.

Homeotic Homeosis describes the transformation of one organ into the identity of another. This can be the result of mutations or miss-expression of homeotic genes; master regulators involved in the patterning of body and organ structures.

Homo-diploid A homo-diploid cell carries two identical copies of the genome. In the majority of species, the central cell is homo-diploid because it contains two nuclei that are derived from the same meiotic product through mitotic divisions. These two nuclei fuse prior to or at fertilization with a haploid sperm to form the triploid endosperm.

Hybrid A hybrid is an offspring resulting from sexual reproduction between two parents belonging to different varieties, species, or even genera. In plants, hybridization is very common and an important driver of evolutionary change. Hybridization can lead to changes at the genomic and epigenomic scale, e.g., the activation of transposition or changes in DNA methylation, respectively.

Hyperplasia Increase in the number of cells in an organ or tissue due to loss of proliferation control. These cells appear normal under a microscope. They are not cancer but may become cancerous.

Hypomorphic It is a mutation that causes a partial loss of gene function. A hypomorph is a reduction in gene function through reduced expression at RNA or protein levels or reduced functional activity, but not a complete loss.

Imaginal disc Imaginal cells are tissue-specific progenitors determined during fly embryogenesis. They increase in size by proliferation during embryonic and larval life, but cells do not differentiate. At metamorphosis the hormone Ecdysone triggers the formation of pupal and adult structures, while the larval tissues are eliminated. The imaginal discs start differentiating giving rise to the external cuticle and appendages, like wing, leg or antennae.

Indels It refers to insertion and/or deletion of nucleotides into genomic DNA and include events less than 1 kb in length.

Insulator Insulators are a class of DNA elements having the ability to protect genes from inappropriate regulatory signals radiating from their neighboring environment. Insulators can control the interaction of enhancers with the promoter. Insulators can also protect genes by acting as barriers preventing the advance of nearby condensed chromatin that might otherwise silence expression (see also PEV).

Interchromatin compartment The interchromatin compartment (IC) refers to the space in the nucleus that is not occupied by chromatin. It consists of channels into chromatin territories that are thought to facilitate the diffusion of transcription factors and messenger RNA processing and transport.

LINE1 Active and autonomous transposable element that propagates in the genome through retrotransposition.

Metastatic cancer It is a cancer that has spread from the part of the body where it started (the primary site) to other parts of the body. When cancer cells break away from a tumor, they can travel to other parts of the body through the bloodstream or the lymph system.

Methylome The methylome describes the entirety of methylated cytosines in the genome of an organism, tissue, or cell type.

Microspore The microspore is the product of meiosis in the male reproductive organs of flowering plants. All four microspores survive and form mature pollen through two mitotic divisions. Each pollen contains two sperm cells that are delivered by the pollen tube to the female gametes to effect double fertilization.

Myeloproliferation It is the abnormal proliferation of myelopoietic cells from bone marrow. Myeloproliferative disorders refer to a type of disease in which the bone marrow produces too many red blood cells, platelets, or granulocytes (a type of white blood cells).

Neoplasm An abnormal mass of tissue that forms when cells grow and divide more than they should or do not die when they should. Neoplasms may be benign (not cancer) or malignant (cancer). Benign neoplasms may grow large but do not spread into, or invade, nearby tissues or other parts of the body. Malignant neoplasms can spread into, or invade, nearby tissues. They can also spread to other parts of the body through the blood and lymph systems.

Nonsense mutations It is a point mutation in a sequence of DNA that results in a premature stop codon, causing a protein to terminate or end its translation earlier than required. In contrast, missense mutation is a point mutation that results in the substitution of one amino acid for another in the protein.

Nucleosome free region Nucleosome free regions refer to a single missing nucleosome at the transcription start site of active genes. It is a consequence of the establishment of DNA accessibility at the promoter, which mediates the binding of transcription factors and assembly of the transcriptional machinery. Nucleosomes in the vicinity of the nucleosome free region are strongly positioned. The +1 and −1 nucleosome are located downstream and upstream of the transcription start site, respectively. The histone variant H2A.Z is found to be incorporated specifically at the +1 nucleosome and several downstream nucleosomes.

Oncogenes Genes that have the potential to become dominant oncogenic determinants in tumorigenesis.

Oncometabolites Metabolites whose quantity increase in tumors compared to normal cells.

Ontogeny This refers to the developmental life-time history of a single organism in comparison to phylogeny which refers to the evolutionary history of a species.

Ovule The ovule is the female reproductive organ of flowering plants and the site of meiosis and female gametogenesis. After meiosis, only one of the meiotic products survives, the functional megaspore, which forms a multicellular organism through several mitotic divisions. Two of the formed cells are the female gametes, the egg and central cells. After fertilization, the ovule develops into the seed.

Paracrine signaling The transfer of information from one cell to another, where the signal molecule travels from the signal-producing cell to the receiving cell by passive diffusion or bulk flow in intercellular fluid. The signaling cell and the receiving cell are usually in the vicinity of each other.

Paused polymerase Pol II pausing or stalling is caused by different incidences; blockage by nucleosomes, need for proofreading, and others. In this context, promoter-proximal pausing is considered an important regulatory stage during transcriptional initiation. At the promoter, stalled Pol II produces short characteristic RNA transcripts. Upon an additional activating cascade of factors (see text) the polymerase is released and produces the productive transcript. Pausing could be required for a better synchrony in gene activation, as a checkpoint for assembly of transcriptional components, a requirement of additional signal integration and others.

Pericentric chromatin/regions/repeats The centromeres of mammalian chromosomes are flanked by repeated sequences that consist of the minor and major satellite repeat sequences. These pericentric repeats are distinct from the centromeres and contribute to a heterochromatic environment that supports centromere function. The heterochromatin that is adjacent to the centromeres is referred to as pericentric heterochromatin.

Phagocytosis Describes the process by which a cell (termed phagocyte) uses its plasma membrane to engulf large particles or other cells giving rise to an internal compartment called phagosome. It is a particular form of endocytosis. Phagocytosis is a major mechanism of the immune system to remove pathogens and cell debris.

Phyla Phyla (singular: phylum) represent the level of classification or taxonomic rank below kingdom and above class.

Pioneer TF Pioneer transcription factor can recognize and bind to their cognate sequence even when the DNA is wrapped in a nucleosomal structure. Pioneer TFs attract nucleosome remodeling complexes and histone modifying enzymes to further loosen and open condensed chromatin for access to additional factors required for example to assemble a transcription initiation complex on free DNA.

Position effect variegation (PEV) Position effect variegation (PEV) causes a gene to be stochastically silenced in regions where it should normally be expressed. This occurs when the gene becomes juxtaposed to heterochromatin in a genome translocation or insertion event. Repressive heterochromatin spreads, un-inhibited by insulator elements, into the regulatory part of the gene. An example is variegated eye color in Drosophila caused by a translocation of the white gene into the vicinity of heterochromatin.

Protamines Protamines are small, arginine-rich proteins replacing the histones in chromatin during late spermatogenesis. They allow a denser packaging of the DNA and are considered crucial for sperm head condensation and genome stabilization.

Prototrophic organism An organism or cell capable of synthesizing all its metabolites from inorganic material, requiring no organic nutrients.

Pseudoautosomal region In heteromorphic sex chromosome systems pairing is restricted to specific regions that share homology. For example, the mammalian X and Y chromosomes have diverged but share small regions of homology at the ends of the chromosomes. Through these regions pairing and exchange of sequences by crossing over occurs during meiosis in the male germline. These pairing regions on the X and Y, thus, behave like autosomes and are referred to as pseudo autosomal regions (PAR) of the sex chromosomes.

Retrovirus element A retrovirus inserts a copy of its RNA genome into the DNA of the host cell it invades. The RNA is copied into DNA by reverse transcriptase encoded by the viral genome. The viral DNA is then integrated and become part of the host genome. There it can be transcribed again, reactivating the process of new viral production. Genomes are full of such retroelements which in most cases represent dead, defective viral genomes.

Sciara/Fungus gnat *Sciara* (gnat), a genus of fungus gnats; insects that have been studied for the process of genome elimination. Analysis of genetics and chromatin in *Sciara* have also revealed insights into the evolution of epigenetic processes. Fungus gnats are harmless and feed on soil-grown fungi. Larvae in some species are plant pathogens and can damage particularly seedlings and young plants.

Sex combs A set of dark bristles found on the first leg pair of Drosophila males and used for mating. Conspicuous morphological marker used by geneticists to identify homeotic transformations of the second and third leg pairs into the identity of the first leg pair.

Synthetic lethality It is the setting in which inactivation of either of two genes individually has little effect on cell viability whereas loss of function of both genes simultaneously leads to cell death. In cancer, the concept of synthetic lethality has been applied to therapy against cancer cells in which one gene is inactivated by deletion or mutation present and the other is pharmacologically inhibited, leading to death of cancer cells. Normal or cancer cells not containing these mutations survive.

Teratocarcinoma Teratocarcinoma is a form of malignant germ cell tumor found in animals and humans. These tumors contain an undifferentiated embryonal carcinoma component and differentiated derivatives that can include all three germ layers.

Transdetermination Transdetermination is defined as a switch in lineage commitment in progenitor cells (i.e. leg imaginal discs) to a developmentally closely related cell type (i.e. wing imaginal disc). This change of fate can occur for example in long-term transplantation experiments of Drosophila larval imaginal discs.

Transdifferentiation Transdifferentiation occurs when a differentiated somatic cell switches to another differentiated lineage without an intermediate stage of dedifferentiation (like a pluripotent or progenitor state).

Transposable element A transposable element (TE) is a DNA sequence capable of changing its position in the genome. The TE codes for enzymes necessary for its own excision and integration. Insertions of TEs can cause mutations and large-scale genome aberrations.

Tumor suppressor gene Class of genes that regulate cell proliferation and their inactivation can play a profound role in the development of malignancies.

X inactivation center The X inactivation center (Xic) refers to a genetic locus on the mouse and human X chromosome that contains regulatory elements for X chromosome inactivation. The Xic includes the Xist gene as well as elements that allow to sense the cell how many X chromosomes are present. All but one X chromosome are inactivated, which results in balanced gene expression of a single X chromosome in both male and female cells.

Xenograft It is one of the most widely used models in cancer research. Typically, human tumor cells are transplanted, either under the skin or into the organ type in which the tumor originated, into immunocompromised mice that do not reject human cells.

Permissions

All chapters in this book were first published by Springer; hereby published with permission under the Creative Commons Attribution License or equivalent. Every chapter published in this book has been scrutinized by our experts. Their significance has been extensively debated. The topics covered herein carry significant information for a comprehensive understanding. They may even be implemented as practical applications or may be referred to as a beginning point for further studies.

The contributors of this book come from diverse backgrounds, making this book a truly international effort. We would like to thank all the contributing authors for lending their expertise to make the book truly unique. They have played a crucial role in the development of this book. Without their invaluable contributions this book wouldn't have been possible. They have made vital efforts to compile up to date information on the varied aspects of this subject to make this book a valuable addition to the collection of many professionals and students.

This book was conceptualized with the vision of imparting up-to-date and integrated information in this field. To ensure the same, a matchless editorial board was set up. Every individual on the board went through rigorous rounds of assessment to prove their worth. After which they invested a large part of their time researching and compiling the most relevant data for our readers.

The editorial board has been involved in producing this book since its inception. They have spent rigorous hours researching and exploring the diverse topics which have resulted in the successful publishing of this book. They have passed on their knowledge of decades through this book. To expedite this challenging task, the publisher supported the team at every step. A small team of assistant editors was also appointed to further simplify the editing procedure and attain best results for the readers.

Apart from the editorial board, the designing team has also invested a significant amount of their time in understanding the subject and creating the most relevant covers. They scrutinized every image to scout for the most suitable representation of the subject and create an appropriate cover for the book.

The publishing team has been an ardent support to the editorial, designing and production team. Their endless efforts to recruit the best for this project, has resulted in the accomplishment of this book. They are a veteran in the field of academics and their pool of knowledge is as vast as their experience in printing. Their expertise and guidance has proved useful at every step. Their uncompromising quality standards have made this book an exceptional effort. Their encouragement from time to time has been an inspiration for everyone.

The publisher and the editorial board hope that this book will prove to be a valuable piece of knowledge for students, practitioners and scholars across the globe.

Index

www.ingramcontent.com/pod-product-compliance
Lightning Source LLC
LaVergne TN
LVHW080603210926
848485LV00003B/96